United States Nuclear Regulatory Commission

*Protecting People and the Environment*

NUREG-1520, Rev. 1

# Standard Review Plan for the Review of a License Application for a Fuel Cycle Facility

# Final Report

Office of Nuclear Material Safety and Safeguards

# AVAILABILITY OF REFERENCE MATERIALS
## IN NRC PUBLICATIONS

United States Nuclear Regulatory Commission

*Protecting People and the Environment*

NUREG-1520, Rev. 1

# Standard Review Plan for the Review of a License Application for a Fuel Cycle Facility

# Final Report

Manuscript Completed: May 2010
Date Published: May 2010

Office of Nuclear Material Safety and Safeguards

# ABSTRACT

NUREG-1520, "Standard Review Plan (SRP) for the Review of a License Application for a Fuel Cycle Facility" (hereafter referred to as the SRP) provides guidance to the staff reviewers in the U.S. Nuclear Regulatory Commission's (NRC's) Office of Nuclear Material Safety and Safeguards who perform safety and environmental impact reviews of applications to construct or modify and operate nuclear fuel cycle facilities. The SRP is intended to be a comprehensive and integrated document that provides the reviewer with guidance that describes methods or approaches that the staff has found acceptable for meeting NRC requirements. As such, this SRP ensures the quality, uniformity, and predictability of the staff reviews. This SRP also makes information about licensing acceptance criteria widely available to interested members of the public and the regulated industry and is intended to improve industry and public stakeholder understanding of the staff review process. Each SRP section addresses the responsibilities of the staff reviewers, the matters that they review the Commission's regulations pertinent to specific technical matters, the acceptance criteria used by the staff, the process and procedures used to accomplish the review, and the conclusions that are appropriate to summarize the review.

This SRP also addresses the long-standing health, safety, and environmental protection requirements of Title 10 of the _Code of Federal Regulations_ (10 CFR) Part 20, "Standards for Protection against Radiation," and 10 CFR Part 70, "Domestic Licensing of Special Nuclear Material," as well as the amended accident safety requirements reflected in 10 CFR Part 70 Subpart H, "Additional Requirements for Certain Licensees Authorized To Possess a Critical Mass of Special Nuclear Material."

Subpart H of 10 CFR Part 70 identifies risk-informed performance requirements and requires applicants and existing licensees to conduct an integrated safety analysis (ISA) and submit an ISA Summary, as well as other information. Chapters 3 (ISA and ISA Summary) and 11 (management measures) of this SRP are the primary chapters that address the staff's review in relation to the performance and other related requirements of Subpart H.

This SRP is not a substitute for NRC regulations and compliance is not required. The approaches and methods in this report are provided for information only. Methods and solutions different from those described in this report will be acceptable if they provide a basis for the staff to make the determination needed to issue or continue a license.

This SRP focuses on safety and environmental impact reviews. The review criteria applicable to the safeguards sections of license applications are published in NUREG -1280, "Standard Format and Content Acceptance Criteria for the Material Control and Accounting (MC&A) Reform Amendment: 10 CFR Part 74, Subpart E," issued April 1995, and NUREG-1065, "Acceptable Standard Format and Content for the Fundamental Nuclear Material Control (FNMC) Plan Required for Low-Enriched Uranium Facilities," issued December 1995.

# CONTENTS

# EXECUTIVE SUMMARY

NUREG-1520, "Standard Review Plan for the Review of a License Application for a Fuel Cycle Facility" (hereafter referred to as the SRP) provides U.S. Nuclear Regulatory Commission (NRC) guidance for reviewing and evaluating the health, safety, and environmental protection aspects of applications for licenses to possess and use special nuclear material (SNM) to produce nuclear reactor fuel. This guidance also applies to the review and evaluation of proposed amendments and license renewal applications for nuclear fuel cycle facilities.

The principal purpose of this SRP is to ensure the quality and uniformity of reviews conducted by the staff of the NRC's Office of Nuclear Material Safety and Safeguards (NMSS). This SRP also provides a well-defined foundation from which to evaluate proposed changes in the scope, level of detail, and acceptance criteria of reviews. Another important purpose of this SRP is to make information about regulatory reviews widely available and to improve communication and understanding of the staff review process. In addition, because this SRP describes the scope, level of detail, and acceptance criteria for reviews, it serves as regulatory guidance for applicants who need to determine what information to present in a license application and related documents.

This SRP addresses the long-standing health, safety, and environmental protection requirements of Title 10 of the *Code of Federal Regulations* (10 CFR) Part 20, "Standards for Protection against Radiation," and 10 CFR Part 70, "Domestic Licensing of Special Nuclear Material," as well as the accident safety requirements reflected in 10 CFR Part 70 Subpart H, "Additional Requirements for Certain Licensees Authorized To Possess a Critical Mass of Special Nuclear Material." For example, the chapters concerning radiation safety, environmental protection, emergency management, and decommissioning contain acceptance criteria that are primarily set by regulations that remain unaffected by the recent revision to 10 CFR Part 70. The review criteria applicable to the safeguards sections of license applications are published in NUREG-1280, "Standard Format and Content Acceptance Criteria for the Material Control and Accounting (MC&A) Reform Amendment: 10 CFR Part 74, Subpart E," issued April 1995 (for high-enriched uranium facilities), and NUREG-1065, "Acceptable Standard Format and Content for the Fundamental Nuclear Material Control (FNMC) Plan Required for Low-Enriched Uranium Facilities," issued December 1995.

Subpart H of 10 CFR Part 70 identifies risk-informed performance requirements and requires applicants and existing licensees to conduct an integrated safety analysis (ISA) and submit an ISA Summary, as well as other information. Chapters 3 (ISA and ISA Summary) and 11 (management measures) of this SRP are the primary chapters that address the staff's review in relation to the performance and other related requirements of Subpart H.

Each nuclear fuel cycle facility license application should contain a safety program description that addresses all of the topics listed in the table of contents of this SRP, in the same order in which they are presented in this SRP. In general terms, the requirements in 10 CFR Part 70 specify the information that an applicant must supply in its safety program description. This SRP complements 10 CFR Part 70 by identifying the specific information that an applicant should submit for staff evaluation.

Separate chapters of this SRP discuss the major topics addressed within the safety program description of a facility license application, including general information, organization and administration, ISA and ISA summary, radiation protection, nuclear criticality safety, chemical process safety, fire safety, emergency management, environmental protection, decommissioning, and management measures. Each of these chapters contains seven sections: (1) purpose of review; (2) responsibility for review; (3) areas of review,

(4) acceptance criteria; (5) review procedures; (6) evaluation findings; and (7) references. Prospective applicants should study the topic areas treated in the chapters of this SRP, paying particular attention to areas of review and acceptance criteria In addition, in accordance with 10 CFR 70.62, "Safety Program and Integrated Safety Analysis," and 10 CFR 70.65, "Additional Content of Applications," the agency requires applicants to submit an ISA Summary in conjunction with the application.

This SRP provides information and guidance to assist the licensing staff and the applicant in understanding the underlying objectives of the regulatory requirements, the relationships among NRC requirements, the licensing process, the major guidance documents that the NRC staff has prepared for licensing fuel cycle facilities, and information about aspects of the staff review process set out in individual SRP sections. Staff analyses are intended to provide regulatory confirmation of reasonable assurance of safe design and operation. A staff determination of reasonable assurance leads to a decision to issue or renew a license or to approve an amendment to the license. If the staff determines that an application contains inadequate information or commitments, the staff will inform the applicant of what is needed and the basis on which the determination was made.

The acceptance criteria delineated in this SRP are intended to communicate the underlying objectives of the NRC's regulations, but they do not represent the only means of satisfying those objectives. Rather an applicant should tailor its safety program to the particular features of its facility. If an applicant chooses approaches other than those presented in this SRP, the applicant should identify the portions of its license application that differ from the design approaches and acceptance criteria of the SRP and should demonstrate how the proposed alternatives provide an acceptable method of complying with the NRC's regulations. The staff retains the responsibility to make an independent determination concerning the adequacy of the applicant's proposed approaches.

# ACRONYMS AND ABBREVIATIONS

| | |
|---|---|
| ACGIH | American Conference of Governmental Industrial Hygienists |
| AEC | active engineered control |
| AEGL | Acute Exposure Guideline Level |
| ALARA | as low as is reasonably achievable |
| ANS | American Nuclear Society |
| ANSI | American National Standards Institute |
| ASME | American Society of Mechanical Engineers |
| ASTM | American Society for Testing and Materials |
| B.A. | bachelor of arts |
| B.S. | bachelor of science |
| BDC | baseline design criterion/criteria |
| CAAS | criticality accident alarm system |
| CFR | *Code of Federal Regulations* |
| CD | chemical dose |
| Ci | curie(s) |
| CM | configuration management |
| DFP | decommissioning funding plan |
| DOE | U.S. Department of Energy |
| DP | decommissioning plan |
| EA | environmental assessment |
| EAL | emergency action level |
| EIS | environmental impact statement |
| ERDA | Energy Research and Development Administration |
| ERPG | Emergency Response Planning Guidelines |

| | |
|---|---|
| FCSS | Division of Fuel Cycle Safety and Safeguards |
| FHA | fire hazards analysis |
| FONSI | finding of no significant impact |
| FR | *Federal Register* |
| GB | gigabecquerels |
| HFE | human factors engineering |
| HPS | Health Physics Society |
| HS&E | health, safety, and environment |
| HIS | human-systems interface |
| ICRP | International Commission on Radiological Protection |
| IEF | initiating event frequency |
| IROFS | item(s) relied on for safety |
| ISA | integrated safety analysis |
| ISO | International Organization for Standardization |
| kg | kilogram |
| km | kilometer |
| MDC | minimum detectable concentration |
| mg | milligram |
| mi | mile |
| MTTF | mean time to failure |
| MTTR | mean time to repair |
| NCS | nuclear criticality safety |
| NCRP | National Council on Radiation Protection and Measurements |
| NFPA | National Fire Protection Association |
| NMSS | Nuclear Material Safety and Safeguards, Office of (NRC) |
| NRC | U.S. Nuclear Regulatory Commission |

| | |
|---|---|
| OCB | oxide conversion building |
| OER | operating experience review |
| ORR | operational readiness review |
| OSHA | Occupational Safety and Health Administration |
| PEC | passive engineered control |
| PFOD | probability of failure on demand |
| PHA | process hazard analysis |
| PM | preventive maintenance |
| PMF | probable maximum flood |
| QA | quality assurance |
| RAI | request for additional information |
| RD | radiological dose |
| RWP | radiation work permits |
| SBC | Standard Building Code |
| SER | safety evaluation report |
| SNM | special nuclear material |
| SRP | standard review plan |
| SSE | safe-shutdown earthquake |
| Sv | sievert |
| TEDE | total effective dose equivalent |
| $UF_6$ | uranium hexafluoride |
| $UO_2$ | uranium dioxide |
| V&V | verification and validation |
| yr | year |

# GLOSSARY

This glossary defines technical/industry terms that are used consistently throughout this SRP, or references the related definitions in either 10 CFR 20.1003 or 10 CFR 70.4, both titled "Definitions." This glossary does not define terms that may have different connotations in different contexts; such terms are defined in the various chapters of this SRP.

| | |
|---|---|
| Active engineered control | A physical device that uses active sensors, electrical components, or moving parts to maintain safe process conditions without any required human action. |
| Accident sequence | An unintended sequence of events that, given the failure of certain items relied on for safety identified in the sequence, would result in environmental contamination, radiation exposure, release of radioactive material, inadvertent nuclear criticality, or exposure to hazardous chemicals (provided that the chemicals are produced from licensed radioactive material). The term "accident" may be used interchangeably with "accident sequence." |
| Acute | This term is defined in 10 CFR 70.4. |
| Administrative control | Either an augmented administrative control or a simple administrative control, as defined herein. |
| Analytical limit | A limit of measured or calculated variables established by the licensee's safety analysis to ensure that safety limits are not exceeded. The safety analysis establishes an analytical limit in terms of a measured or calculated variable and a specific time after the value is reached to begin protective action. The analysis should account for the dynamic and transient nature of certain process variables and ensures these variables do not exceed the safety limit as a result of this transient behavior. |
| Augmented administrative control | A procedurally required or prohibited human action, combined with a physical device that alerts the operator that the action is needed to maintain safe process conditions or that otherwise adds substantial assurance of the required human performance. |
| Available and reliable to perform their function when needed | This term is defined in 10 CFR 70.4. |
| Baseline design criteria | A set of criteria specifying design features and management measures that are required and acceptable under certain |

conditions for new processes or facilities specified in 10 CFR 70.64, "Requirements for New Facilities or New Processes at Existing Facilities." In general, these criteria are the acceptance criteria that apply to safety design for new facilities and new processes, as described in this SRP.

Configuration management

This term is defined in 10 CFR 70.4.

Controlled area

This term is defined in 10 CFR 20.1003.

Controlled parameter

A measurable parameter that is maintained within a specified range by one or more specific controls to ensure the safety of an operation.

Consequence

Any result of interest caused by an event or sequence of events. In this context, "adverse consequence" refers to adverse health or safety effects on workers, the public, or the environment.

Critical mass of special nuclear material

This term is defined in 10 CFR 70.4.

Double contingency protection

A characteristic or attribute of a process that has incorporated sufficient safety factors so that at least two unlikely, independent, and concurrent changes in process conditions are required before a nuclear criticality accident is possible.

Engineered control

See "active engineered control" and "passive engineered control."

External event

An event for which the likelihood cannot be altered by changes to the regulated facility or its operation. This would include all natural phenomena events, plus airplane crashes, explosions, toxic releases, fires, etc., occurring near or on the plant site.

Hazardous chemicals produced from licensed materials

This term is defined in 10 CFR 70.4.

Integrated safety analysis

This term is defined in 10 CFR 70.4.

Integrated safety analysis summary

This term is defined in 10 CFR 70.4.

Items relied on for safety

This item is defined in 10 CFR 70.4 and includes all safety

controls, as defined in this SRP.

| | |
|---|---|
| Management measures | This term is defined in 10 CFR 70.4. |
| Mitigative control | A control intended to reduce the consequences of an accident sequence, not to prevent it. When a mitigative control works as intended, the results of the sequence are called the mitigated consequences. |
| Natural phenomena event | Earthquakes, floods, tornadoes, tsunamis, hurricanes, and other events that occur in the natural environment and could adversely affect safety. Natural phenomena events may be credible or incredible, depending on their likelihood of occurrence. |
| New processes at existing facilities | Systems-level or facility-level design changes to process equipment, process technology, facility layout, or types of licensed material possessed or used. Generally, this definition does not include component-level design changes or equipment replacement. |
| Operating limit | A limiting value (or range of values) for a process parameter at which the plant operators normally operate the facility. |
| Passive engineered control | A device that uses only fixed physical design features to maintain safe process conditions without any required human action. |
| Preventive control | A control intended to prevent an accident (i.e., any of the radiological or chemical consequences described in 10 CFR 70.61, "Performance Requirements"). |
| Safety control | A system, device, or procedure that is intended to regulate a device, process, or human activity to maintain a safe state. Controls may be engineered controls or administrative (procedural) controls, and they may be either preventive or mitigative, as defined herein. |
| Safety limit | A limit chosen to maintain the integrity of physical barriers that protect against exceeding the performance requirements of 10 CFR 70.61. |
| Safe process conditions | The defined ranges or sets of acceptable values of one or more controlled parameters. |
| Setpoint | A predetermined value for actuation of the final setpoint device to initiate a protective action. |

Simple administrative control          A procedural human action that is prohibited or required to
                                       maintain safe process conditions.

Unacceptable performance               This term is defined in 10 CFR 70.4.
deficiencies

Worker                                 This term is defined in 10 CFR 70.4.

# INTRODUCTION

NUREG-1520, "Standard Review Plan (SRP) for the Review of a License Application for a Fuel Cycle Facility" (hereafter referred to as the SRP) provides U.S. Nuclear Regulatory Commission (NRC) guidance for reviewing and evaluating the health, safety, and environmental protection aspects of applications for licenses to possess and use special nuclear material (SNM) to produce nuclear reactor fuel. This guidance is specific to fuel cycle facilities regulated under Title 10 of the _Code of Federal Regulations_ (10 CFR) Part 70, "Domestic Licensing of Special Nuclear Material," that is, facilities that are authorized for or are seeking a license to possess and use more than a critical mass of SNM. This guidance also applies to the review and evaluation of proposed amendments and license renewal applications for nuclear fuel cycle facilities. This guidance does not apply to conversion facilities,[1] gaseous diffusion plants,[2] reprocessing facilities, and plutonium processing facilities.[3]

The principal purpose of this SRP is to ensure the quality and uniformity of reviews conducted by the staff of the NRC's Office of Nuclear Material Safety and Safeguards (NMSS). This SRP also provides a well-defined foundation from which to evaluate proposed changes in the scope, level of detail, and acceptance criteria of reviews. Another important purpose of this SRP is to make information about regulatory reviews widely available and to improve communication and understanding of the staff review process. In addition, because this SRP describes the scope, level of detail, and acceptance criteria for reviews, it serves as regulatory guidance for applicants who need to determine what information to present in a license application and related documents.

This SRP addresses the long-standing health, safety, and environmental protection requirements of 10 CFR Part 20, "Standards for Protection against Radiation," and 10 CFR Part 70, as well as the accident safety requirements reflected in 10 CFR Part 70 Subpart H, "Additional Requirements for Certain Licensees Authorized To Possess a Critical Mass of Special Nuclear Material." The NRC review criteria applicable to the safeguards sections of license applications are published in NUREG1280, "Standard Format and Content Acceptance Criteria for the Material Control and Accounting (MC&A) Reform Amendment: 10 CFR Part 74, Subpart E," issued April 1995 (for high-enriched uranium facilities), and NUREG-1065, "Acceptable Standard Format and Content for the Fundamental Nuclear Material Control Plans for high enriched uranium facilities and low enriched uranium facilities, respectively(FNMC) Plan Required for Low-Enriched Uranium Facilities," issued December 1995.

Subpart H of 10 CFR Part 70 identifies risk-informed performance requirements and requires applicants and existing licensees to conduct an integrated safety analysis (ISA) and submit an ISA Summary, as well as other information. Chapters 3 (ISA and ISA Summary) and 11

---

[1] The NRC regulates conversion facilities under the provisions of 10 CFR Part 40, "Domestic Licensing of Source Material."
[2] The NRC regulates gaseous diffusion plants under 10 CFR Part 76, "Certification of Gaseous Diffusion Plants." This regulation specifically applies to those portions of the Portsmouth and Paducah Gaseous Diffusion Plants located in Piketon, OH, and Paducah, KY, respectively, that are leased by the United States Enrichment Corporation.
[3] Guidance for the review of a license application for a Mixed Oxide (MOX) Fuel Fabrication Facility is provided in NUREG-1718, "Standard Review Plan for the Review of a License Application for a MOX Fuel Fabrication Facility," issued August 2000.

(Management Measures) of this SRP are the primary chapters that address the staff's review in relation to the performance and other related requirements of Subpart H. For new facilities that have not already been designed, built, licensed and operated, Subpart H also requires adherence to baseline design criteria, as specified in 10 CFR 70.64, "Requirements for New Facilities or New Processes at Existing Facilities."

This SRP is a guidance document that is intended for use during the review of license applications, license renewal applications, and amendment applications. This SRP does not preclude licensees or applicants from suggesting alternative approaches to those specified in the SRP to demonstrate compliance with applicable regulations.

In reviewing a license application, renewal application, or license amendment for a fuel cycle facility, the staff must determine whether there is reasonable assurance that the facility can and will be operated in a manner that will not be inimical to the common defense and security, and will adequately protect the health and safety of workers, the public, and the environment. To carry out this responsibility, the staff evaluates the information that the applicant provides and, through independent assessments, determines whether the applicant has proposed an adequate safety program that is compliant with regulatory requirements. To assist the staff in carrying out this responsibility, this SRP clearly states and identifies those standards, criteria, and bases that the staff will use in reaching licensing decisions.

An application for a 10 CFR Part 70 license must include specific information on the proposed equipment and facility in accordance with 10 CFR 70.22(a)(7), which states that each application shall contain the following:

> A description of equipment and facilities which will be used by the applicant to protect health and minimize danger to life or property (such as handling devices, working areas, shields, measuring and monitoring instruments, devices for the disposal of radioactive effluents and wastes, storage facilities, criticality accident alarm systems, etc.).

In reviewing 10 CFR Part 70 license applications, the staff uses a reasonable assurance paradigm and focuses on the programmatic provisions of the applicant's proposed activities. Consequently, the licensing decision is ultimately based on information with a sufficient level of detail that permits reviewers to understand process system functions, and functionally, how items relied on for safety (IROFS) can perform as intended and be reliable. This staff review method is intended to ensure that the staff decision is based on a reasonable assurance that the submitted ISA Summary is complete and that the licensee will comply with the ISA and maintain it consistent with the regulations. The level of detail required for a licensing decision generally does not require a final facility design; however, identification of all IROFS and possible accident sequences is necessary to make a licensing decision. Even though detailed information about each IROFS is not required, sufficient information has to be provided to understand the process, theory of operation and functions of each IROFS and reasonable assurance that the ISA Summary is complete. For uranium enrichment facilities, to ensure that the applicant's programs have been sufficiently implemented and commitments have been properly applied in the final facility design and in the constructed facility, 10 CFR 70.32(k) states that the following:

"No person may commence operation of a uranium enrichment facility until the Commission verifies through inspection that the facility has been constructed in accordance with the requirements of the license."

This requirement applied through inspections, and not by licensing reviews, will ensure that the programmatic commitments made by licensee are properly applied in the as built facility. This inspection is intended to inspect the final design of the facility and the procedures that have been prepared to implement the licensee's commitments that are reflected in the license. The purpose of the review is to verify through inspection that the facility has been constructed in accordance with its license. Furthermore, for significant modifications to existing fuel cycle facilities, such as the licensing and construction of new processes, the staff may impose a license condition that specifies that an operational readiness review (ORR) inspection be conducted before operation to verify that the new part of the facility has been constructed in accordance with the requirements of the license. To facilitate the planning and accomplishment of a risk-informed ORR, the staff relies upon the licensee to provide a complete set of information. This complete set of information has been referred to in some projects as IROFS boundary packages.[4] For simplicity they will be referred to hereinafter in this document as IROFS boundary packages. Regardless of what they are called in a license application, the key point is that they provide information to the reviewers and inspectors about supporting systems that directly affect the effectiveness of the IROFS and the reliability and availability of the IROFS as required by 10 CFR 70.62(d). Inspectors use this information during the ORR inspection to determine if the licensee meets the requirements in 10 CFR 70.23(a)(3)–(4) and in 10 CFR 70.61(e).

In developing the performance requirements in 10 CFR Part 70, the NRC anticipated that, in the future, changes will be made to the facility design and processes and, therefore, described a process for addressing these changes is described in 10 CFR 70.72, "Facility Changes and Change Processes.". For a uranium enrichment facility, the licensee may make changes to its design, after receiving its license, during the construction phase and after operations begin.

---

[4] IROFS boundary packages are documents that contain the physical descriptions and parameters of structures, systems, and components that are used to meet the performance requirements of 10 CFR 70.61, "Performance Requirements." IROFS boundary definition packages are also prepared for administrative procedures or worker actions that are defined as IROFS. The boundary packages identify the specific functions to be performed by an IROFS and identify any items that may affect the function of the IROFS. The boundary packages also identify the facility areas in which the IROFS is used, design and functional attributes, management measures, any open items, and supporting documentation (e.g., piping and instrumentation diagrams, schematics).

Design and functional attributes should include safety functions such as separation from other IROFS, redundancy and diversity, fail-safe design, set points, environmental qualification, seismic qualification, and fire protection. System interfaces such as instrumentation, electrical, cooling, and lubrication requirements should also be included under design and functional attributes.

Management measures should address all of the management measures required to be applied to IROFS under 10 CFR 70.4, "Definitions," and include summary descriptions; references to maintenance, training, and procedures documents; or both, as appropriate for the IROFS. The references should be adequate to identify the actual working-level training or procedures document.

Open items that affect the reliability, the effectiveness, or both of the IROFS should be closed by the time of the ORR. The open items section should identify open items associated with the IROFS during the review and describe how the open items were resolved.

These changes, therefore, need to be submitted and reviewed in accordance with 10 CFR 70.72.

An applicant submits a complete description of the safety program for the possession and use of SNM to show how it will ensure compliance with the applicable requirements. It must describe the safety program in sufficient detail to permit the staff to determine with reasonable assurance that the facility is designed and will be operated without undue risk to the health and safety of workers or the public. Before submitting a program description, an applicant should have analyzed the facility in sufficient detail to conclude that it is designed and can be operated safely.

The requirements in 10 CFR 70.22, "Contents of Applications"; 10 CFR 70.23, "Requirements for the Approval of Applications"; and Subpart H to 10 CFR Part 70 specify, in general terms, the information to be supplied in a safety program description. As such, this SRP identifies the specific information that an applicant should submit for staff evaluation. Prospective applicants should study the topic areas treated in this SRP and the sections within each chapter (particularly those regarding areas of review and acceptance criteria). To facilitate the staff's review, a license application should contain a safety program description that addresses the contents of this SRP in the same order as presented in this document. Applicants may reference material submitted in one location in a license application at another location to avoid unnecessary duplication.

In addition, 10 CFR 70.61 requires each applicant to evaluate, in an ISA performed in accordance with 10 CFR 70.62, "Safety Program and Integrated Safety Analysis," compliance with the performance requirements in 10 CFR 70.61(b), 10 CFR 70.61(c), and 10 CFR 70.61(d). The regulations in 10 CFR 70.65 describe the requirements for the contents of the ISA Summary that must be submitted with the application. According to 10 CFR 70.65(b)(3), the ISA must contain the following:

> A description of each process (defined as a single reasonably simple integrated unit operation within an overall production line) analyzed in the integrated safety analysis in sufficient detail to understand the theory of operation; and, for each process, the hazards that were identified in the integrated safety analysis pursuant to §70.62(c)(1)(i)- (iii) and a general description of the types of accident sequences.

The regulations in 10 CFR 70.65(b)(6) require that the ISA contain the following:

> A list briefly describing each item relied on for safety which is identified pursuant to §70.61(e) in sufficient detail to understand their functions in relation to the performance requirements of §70.61.

Based on the information in the ISA Summary provided in accordance with 10 CFR 70.65, the NRC makes licensing decisions as required under 10 CFR 70.21, "Filing," 10 CFR 70.22, 10 CFR 70.23, and 10 CFR 70.60, "Applicability," through 10 CFR 70.66, "Additional Requirements for Approval of License Application." These decisions include compliance with the performance requirements, the baseline design criteria, defense in depth, and the adequacy of management measures.

This SRP provides information and guidance to assist the licensing staff and the applicant in understanding the underlying objectives of the regulatory requirements, the relationships among NRC requirements, the licensing process, the major guidance documents that the NRC staff has prepared for licensing fuel cycle facilities, and information about aspects of the staff review process set out in individual SRP sections. Staff analyses are intended to provide regulatory confirmation of reasonable assurance of safe design and operation. A staff determination of reasonable assurance leads to a decision to issue or renew a license or to approve an amendment. If the staff determines that an application contains inadequate descriptions or commitments, the staff will inform the applicant of what is needed and the basis on which the determination was made.

The acceptance criteria delineated in this SRP are intended to communicate the underlying objectives, but they do not represent the only means of satisfying those objectives. An applicant should tailor its safety program to the particular features of its facility. If an applicant chooses approaches other than those presented in this SRP, the applicant should identify the portions of its license application that differ from the design approaches and acceptance criteria of the SRP, and should document how the proposed alternatives provide an acceptable method of complying with the Commission's regulations. The staff retains the responsibility to make an independent determination concerning the adequacy of the applicant's proposed approaches.

Each SRP chapter is structured to include the (1) purpose of the review, (2) responsibility for the review, (3) areas of review, (4) acceptance criteria, (5) procedure procedures, (6)) evaluation findings, and (7) ) references.

Purpose of Review

This section presents a brief statement of the purpose and objectives of reviewing the subject areas. It emphasizes the staff's evaluation of the ways the applicant will achieve identified performance objectives and ensures (through the review) that the applicant has used a multi-disciplinary, systems-oriented approach to establish designs, controls, and procedures within individual technical areas.

Responsibility for Review

This section identifies the NRC organization and individuals (by function) who are responsible for evaluating the specific subject or functional area. If reviewers with expertise in other areas are to participate in the evaluation, they also are identified by function. In general, the licensing project manager has responsibility for the total review product, which is referred to as a safety evaluation report (SER). However, an identified technical specialist will have primary responsibility for a particular review topic (usually an SRP chapter), and one or more specialists may have supporting responsibility. This team of specialist reviewers performs the overall application review. Although they individually perform their review tasks, the reviews are extensively coordinated and integrated to ensure consistency in approach and to promote risk-informed reviews. The licensing project manager oversees and directs the coordination of the reviewers. The reviewers' immediate line management has the responsibility to ensure that qualified reviewers perform an adequate review.

Areas of Review

This section describes the topics, functions, systems, components, analyses, applicant commitments, data, or other information that should be reviewed as part of the given subject area of the license application. Because this section identifies information to be reviewed in evaluating the adequacy of the application, it identifies the acceptable content of an applicant's submittal in the areas discussed. The areas of review identified in this section obviate the need for a separate standard format and content guide.

The topics identified in this section also set the content of the next two sections of the SRP, covering the acceptance criteria and review procedures. Applications should address, in the same order, the topics set forth as areas of review. This section also identifies the information needed or the review expected from other NRC staff to permit the individual charged with primary review responsibility to complete the review.

Acceptance Criteria

This section defines a set of applicable NRC acceptance criteria on the basis of regulatory requirements, and these collectively establish the basis for assessing the acceptability of the applicant's commitments relative to the design, programs, or functions within the scope of the particular SRP section. Technical bases consist of specific criteria, such as NRC regulations, regulatory guides, NUREG reports, and industry codes and standards. As such, the acceptance criteria present positions and approaches that are acceptable to the staff. As noted above, the NRC does not consider them to be the only acceptable positions or approaches, and the applicant may propose others.

The requirements for approval of an application appear in 10 CFR 70.23(a). These requirements state that the NRC will approve an application upon finding that the applicant is qualified, the proposed equipment and facilities are adequate to protect health and minimize danger to life or property, and the proposed procedures are adequate. As a technical matter, NMSS will determine how final the design must be to make this finding. The NRC staff will interpret applicant commitments to follow an industry standard as a commitment to adhere to all "shall" statements in the standard. The staff will not consider suggestions and recommendations in the standards (so-called "should" statements) as binding commitments by the applicant, unless the applicant specifically states an intent to treat the "should" statements as binding commitments (i.e., treat them as if they are "shall" statements). The applicant may make such commitments as part of its description of the safety program basis. If the staff finds that a definitive commitment to a "should" statement is necessary to provide adequate protection, the reviewer will raise this as an issue in any request for additional information on specific licensing actions. However, applicants should note that some industry or consensus standards specifically direct users to provide justifications for not abiding by recommendations contained in the standards. For example, American National Standards Institute/American Nuclear Society Standard 8.1, "Nuclear Criticality Safety in Operations with Fissionable Materials Outside Reactors," states that "when recommendations are not implemented, justification shall be provided," thus effectively mixing "should" and "shall" statements. In such instances, applicants should be prepared to justify any decisions not to abide by recommendations contained in the standards.

This SRP presents acceptance criteria for each technical function area (e.g., nuclear criticality safety, fire safety, radiation safety) and the management measures (e.g., configuration management, maintenance, audits, and assessments) that an applicant uses to provide a level of protection commensurate with the accident risk inherent in the proposed process activities. For example, at process stations (or for an entire process or sub-process) for which the inherent risk to workers, the public, or the environment is demonstrably small, the applicant needs to provide only those design and operating controls that ensure that risk remains small. The key element in the staff's evaluation is the applicant's adequate demonstration of an acceptable control of risk, which then supports a competent and informed review by the NRC staff.

Review Procedures

This section describes how the staff will perform the review. It generally describes procedures that the reviewer should follow to achieve an acceptable scope and depth of review and to obtain reasonable assurance that the applicant has provided appropriate commitments to ensure that it will operate the facility safely. This could include identifying which licensee commitments the reviewer needs to verify, and could include directing the reviewer to coordinate with others having review responsibilities for other portions of the application than those assigned to the reviewer. This section should provide whatever procedural guidance is necessary to evaluate the applicant's level of achievement of the acceptance criteria.

Evaluation Findings

This section presents the type of positive conclusion that is sought, for the particular review area, to support a decision to grant a license or amendment. The review must be adequate to permit the reviewer to support this conclusion. For each section, the staff SER publishing the results of the review will include a conclusion of this type. The SER will also contain a description of the review, including aspects that received special emphasis, matters that the applicant modified during the review, matters that require additional information or will be resolved in the future, aspects where the facility's design or the applicant's proposals deviate from the criteria in the SRP, and the bases for any deviations from the SRP or proposed exemptions from the regulations.

In the SER, the staff may recommend license conditions to address any issues that were not previously resolved by an applicant's commitments. Such conditions are discussed with an applicant before issuing the license (or license amendment) and become commitments to performance in addition to those commitments that the applicant presented in the application.

References

This section lists references that the staff should consult during the review process. However, depending on the action and approaches proposed by the applicant, they may not always be relevant to the review.

# 1. GENERAL INFORMATION

## 1.1 Facility and Process Overview

### 1.1.1 Purpose of Review

The purpose of this review is to ascertain whether an application for a new, renewed, or amended license includes an overview of the facility layout and a summary description of its manufacturing processes. All reviewers, U.S. Nuclear Regulatory Commission (NRC) managers, and the public may use this overview to gain a general understanding of the purpose of the facility and an overview of the design of its processes. The integrated safety analysis (ISA) summary describes the facility and its manufacturing processes in more detail.

### 1.1.2 Responsibility for Review

Primary:     Licensing Project Manager

Secondary:   ISA Reviewer, Environmental Reviewer, Emergency Protection Reviewer, other Technical Reviewers

Supporting:  None

### 1.1.3 Areas of Review

The staff should review the general facility and process descriptions provided in the license application. The areas of review should include the following:

- Facility Layout Description—This area includes a description of the purpose of each feature and the interrelationships between features.

- Process Overview—The process description should be a narrative description of the different processes at the facility involving licensed material.

- Site Overview—The description includes the proximity of facility buildings to the site boundary and nearby populations, including the most recent census data available when the license application was submitted.

- Descriptive Summary of Licensed Material—The summary should include the name, amount, and specification (including chemical and physical forms) of the special nuclear material (SNM). The license application also should include a list of raw materials, byproducts, wastes, moderators, and finished products of the facility.

The facility and process description in the license application must be consistent with the information presented in the ISA Summary and Chapter 8, which addresses emergency management, of this Standard Review Plan (SRP).

<u>Review Interfaces</u>

In addition to the general information in the application, the reviewer should examine information in the following areas to ensure that it is consistent with the general information section in the license application:

- information about the facility and site and the different processes that will involve SNM in Chapter 3 of this SRP

- the facility and process descriptions in Chapter 8 of this SRP

## 1.1.4 Acceptance Criteria

### 1.1.4.1 Regulatory Requirements

The regulations applicable to the areas of review in this SRP are Title 10 of the *Code of Federal Regulations* (10 CFR) 70.22, "Contents of Applications," and 10 CFR 70.65(b)(1)–(2).

### 1.1.4.2 Regulatory Guidance

No regulatory guides apply to a general facility description for a fuel cycle facility.

### 1.1.4.3 Regulatory Acceptance Criteria

The reviewer should find the applicant's general information acceptable if it provides reasonable assurance that the acceptance criteria presented below are adequately addressed and satisfied.

1.1.4.3.1 Facility Layout Description

The applicant's overview of the facility is acceptable if it meets the following conditions:

- The application presents information at a level of detail that is appropriate for general familiarization and understanding of the proposed facility. This information should be consistent with that presented in the ISA Summary but may be less detailed.

- The overview should describe the relationship of specific facility features to the major processes that will be ongoing at the facility.

- This description should include the building locations of major process components; drawings illustrating the layout of the buildings and structures within the controlled area boundary should be used to support the description.

- If applicable, the applicant has marked portions of the application to identify any proprietary or sensitive information related to the facility (e.g., the location of certain enrichment processes).

### 1.1.4.3.2  Process Overview

The process overview is acceptable if it summarizes the major chemical or mechanical processes involving licensable quantities of radioactive material based, in part, on information presented in the ISA Summary.  This description should include the building locations of major process components and brief accounts of the process steps.

### 1.1.4.3.3  Site Overview

The license application summarizes the site information contained in the ISA Summary.  This includes descriptions of the overall facility layout and the drawings to support such descriptions.  The license application describes the site's geographical characteristics and facility structural features (such as buildings, towers, and tanks), transportation rights of way, and proximity to nearby populations.  The license application fully describes the facility location.  These descriptions are consistent with the information in Chapter 8 of this SRP.

If applicable, the applicant has portion-marked the application to identify any proprietary or sensitive information (e.g., the location of the controlled area boundary).

### 1.1.4.3.4  Descriptive Summary of Licensed Material

The summary is acceptable if it includes the following:

- The summary should describe chemical and physical forms of SNM in process; the maximum amounts of SNM in process in various building locations; and the types, amounts, and discharge points of waste materials discharged to the environment from the processes.

- The application presents a summary identification of the raw materials, byproducts, wastes, and finished products of the facility.  This information should include data on expected levels of trace impurities or contaminants (particularly fission products or transuranic elements) characterized by identity and concentration.  In addition, this summary should identify the proposed possession at the facility of any moderator or reflector with special characteristics, such as beryllium or graphite.

- If applicable, the applicant has marked portions of the application to identify any proprietary or sensitive information (e.g., possession limits).

## 1.1.5  Review Procedures

### 1.1.5.1  Acceptance Review

During the NRC staff's acceptance review, the staff should screen the submittals to identify major deficiencies in the information provided in each review area in Section 1.1.3.  Reviewers must decide whether they have enough information to proceed with a detailed technical review.  If the reviewer identifies significant deficiencies in the license application during the acceptance review, the staff should ask the applicant to submit additional material or revise the license application.  This will ensure that the license application is technically sufficient before the start of the safety evaluation.  Once the deficiencies in the license application have been addressed during the acceptance review, the NRC reviewer should examine the facility and process descriptions and assess their acceptability by comparison with the acceptance criteria in Section 1.1.4.3 and consistency with information in the ISA Summary and the applicant's emergency management plan.

### 1.1.5.2 Safety Evaluation

The information submitted by the applicant in this section is informational in nature, and no technical analysis is required. In addition, the reviewers use the information in this section only as background for the more detailed descriptions in later sections of the application. Therefore, the primary reviewer ascertains whether the descriptive information presented is consistent with the information presented in the ISA Summary and the emergency management plan.

### 1.1.6 Evaluation Findings

If the license application provides sufficient information and the regulatory acceptance criteria in Section 1.1.4.3 are appropriately satisfied, the staff concludes that this evaluation is complete. The reviewer writes material suitable for inclusion in the safety evaluation report prepared for the entire application. The report includes a statement summarizing what was reviewed and why the reviewer finds the submittal acceptable. The staff can document the review as follows:

> The staff has reviewed the general facility description for [name of facility] according to Section 1.1 of the Standard Review Plan. [Name of facility] has adequately described (1) the facility and its processes so that the staff has an overall understanding of the relationships of the facility features and (2) the function of each feature. [Name of facility] has cross-referenced its general description with the more detailed descriptions elsewhere in the application. The staff also confirmed that the information provided in the license application is consistent with the ISA Summary and the emergency management plan. Therefore, the NRC staff concludes that [Name of facility] has complied with the general requirements of 10 CFR 70.22, "Contents of Applications," and 10 CFR 70.65(b)(1–2), as applicable to this section.

## 1.2 Institutional Information

### 1.2.1 Purpose of Review

The purpose of this review is to establish whether the license application includes adequate information identifying the applicant, the applicant's characteristics, and the proposed activity.

### 1.2.2 Responsibility for Review

Primary:     Licensing Project Manager

Secondary:   None

Supporting:  Office of the General Counsel
Office of Federal and State Materials and Environmental Management
    Programs/Division of Waste Management and Environmental Protection
Office of Nuclear Reactor Regulation/Division of Policy and Rulemaking

### 1.2.3 Areas of Review

The staff should review the institutional information provided by the applicant or licensee in the license application. Information provided for review should include the following:

- Corporate Identity and Ownership—This section should include the identity and physical address of the applicant's facility and corporate headquarters, corporate information sufficient to show the relationship of the applicant's organization relative to other corporate entities, and the existence and extent of foreign ownership or influence.

- Financial Qualifications—Information provided for review in this section should include the applicant's financial qualifications to pursue the activities for which the license is sought.

- Characteristics of the Material—This information should include the type, quantity, and form(s) of material(s) proposed for use at the licensed facility.

- Authorized Uses—The application should clearly describe each proposed licensed activity in the form of requested authorized uses and the type of license the applicant is requesting.

- Special Exemptions or Special Authorizations—The application should clearly describe any special exemptions and authorizations the applicant is requesting and the regulatory requirements for which the applicant is seeking approval or exemption.

- Security of Classified Information—The license application will include this section only if the applicant or licensee has requested and received a facility security clearance in accordance with 10 CFR Part 95, "Facility Security Clearance and Safeguarding of National Security Information and Restricted Data."

- <u>Period of Time for Which the License Is Requested</u>—The license application should specify the period of time for which the applicant is seeking approval.

<u>Review Interfaces</u>

None

## 1.2.4 Acceptance Criteria

### 1.2.4.1 Regulatory Requirements

The regulations applicable to the areas of review in this SRP are 10 CFR 70.22; 10 CFR 70.23, "Requirements for the Approval of Applications"; 10 CFR 70.33, "Renewal of Licenses"; and 10 CFR Part 95.

### 1.2.4.2 Regulatory Guidance

No regulatory guides apply to institutional information for a fuel cycle facility.

### 1.2.4.3 Regulatory Acceptance Criteria

The application is acceptable if it meets the criteria below.

#### 1.2.4.3.1 Corporate Identity

The applicant has furnished its full name and physical address and the address of the fuel cycle facility if it is different from that of the applicant. If the application is for license renewal, the applicant has identified the number of the license to be renewed. The application indicates the State where the applicant is incorporated or organized and the location of the principal office. The application should include any information known to the applicant concerning the control or ownership, if any, exercised over the applicant by any alien, foreign corporation, or foreign government. Primary ownership and relationships to other components of the same ownership are explicitly described. The presence and operations of any other company on the site to be licensed are fully described.

#### 1.2.4.3.2 Financial Qualifications

A description of financial qualifications demonstrates the applicant's current and continuing access to the financial resources necessary to engage in the proposed activity in accordance with 10 CFR 70.22(a)(8) and 10 CFR 70.23(a)(5). For new facilities, the description includes sufficient details demonstrating that the applicant has adequate financial resources to support the safe siting, construction, operation, maintenance, and eventual decommissioning of the proposed facility.

#### 1.2.4.3.3 Characteristics of the Material

The application identifies the elemental name, maximum quantity, and specifications, including the chemical and physical form(s), of the licensable material that the applicant proposes to acquire, deliver, receive, possess, produce, use, transfer, or store. For each SNM, the specifications include the isotopic content and amount of enrichment by weight percent.

### 1.2.4.3.4 Authorized Uses

The application includes a summary, nontechnical narrative description for each activity or process in which the applicant proposes to acquire, deliver, receive, possess, produce, use, process, transfer, or store SNM. The authorized uses of SNM proposed for the facility are described and are consistent with the Atomic Energy Act of 1954, as amended. The description is consistent with more detailed process descriptions submitted as part of the ISA Summary reviewed in Chapter 3 of this SRP.

If the application is for a license renewal, the applicant has clearly stated the time period for which renewal is sought.

### 1.2.4.3.5 Special Exemptions or Special Authorizations

The license application clearly describes any exemptions or authorizations of an unusual nature and adequately justifies them for the NRC's consideration. The applicant has explained any cross-reference to other sections in the license application supporting such justifications. The license application clearly discusses special authorizations and/or exemptions already granted by the NRC.

### 1.2.4.3.6 Security of Classified Information

If applicable, the applicant has requested and received a facility security clearance in accordance with 10 CFR Part 95.

## 1.2.5 Review Procedures

### 1.2.5.1 Acceptance Review

The primary reviewer should determine whether the license application is complete and addresses each issue in Section 1.2.3. If the reviewer identifies significant deficiencies in the license application during the acceptance review, the staff should ask the applicant to submit additional material or revise the license application. This will ensure that the license application is technically sufficient before the start of the safety evaluation. Once the deficiencies in the license application have been addressed during the acceptance review, the NRC reviewer should examine the institutional information and assess its acceptability by comparison with the acceptance criteria in Section 1.2.4.3.

### 1.2.5.2 Safety Evaluation

The information submitted by the applicant in this section is, for the most part, informational in nature, and detailed technical analysis is generally not required beyond the acceptance criteria. For new facilities, the reviewer requests review assistance, as needed, from the Office of the General Counsel, the Office of Federal and State Materials and Environmental Management

Programs/Division of Waste Management and Environmental Protection, and the Office of Nuclear Reactor Regulation/Division of Policy and Rulemaking in the review of corporate and financial information.

## 1.2.6 Evaluation Findings

If the information provided by the applicant is consistent with the guidance of this SRP, the staff will conclude that this evaluation is complete. The staff can document its review as follows:

> The staff has reviewed the institutional information for [name of facility] according to Section 1.2.of the Standard Review Plan. On the basis of the review, the NRC staff has determined that [name of facility] has adequately described and documented the corporate structure and financial information, and is in compliance with those requirements in 10 CFR 70.22 and 10 CFR 70.65 related to other institutional information. In addition, in accordance with 10 CFR 70.22(a)(2), (3), and (4), [name of facility] has adequately described: (1) the period of time for which the license is requested, (2) types, (3) forms, (4) quantities, and (5) the proposed authorized uses of licensed materials to be permitted at this facility as follows:

> Material Form Quantity Authorized Use(s)

> [name of facility]'s proposed activities are consistent with the Atomic Energy Act of 1954, as amended. [Name of facility] has provided all institutional information necessary to understand the ownership, financial qualifications, location of the facility to be licensed, planned activities, and nuclear materials to be handled in connection with the requested license.

## 1.3 Site Description

### 1.3.1 Purpose of Review

The purpose of this review is to determine whether the information provided by an applicant adequately describes the geographic, demographic, meteorological, hydrologic, geologic, and seismologic characteristics of the site and the surrounding area. The site description is a summary of the information that the applicant used in preparing the environmental report, emergency plan, and ISA Summary.

### 1.3.2 Responsibility for Review

Primary: Licensing Project Manager

Secondary: ISA Reviewer, Environmental Protection Reviewer, Emergency Plan Reviewer

Supporting: None

### 1.3.3 Areas of Review

The information in this section of the application is summarized from the information presented in more detail in the applicant's environmental report, emergency management plan, and ISA Summary. The information that the NRC staff will review includes the following (as appropriate for the facility being reviewed):

- site geography

  - site location: State, county, municipality, topographic quadrangle (eight 7-1/2-minute quadrants), site boundary, and controlled area boundary

  - major nearby highways

  - nearby bodies of water

  - any other significant geographic feature that may affect accident analysis within 1.6 kilometers (1 mile) of the site (e.g., ridges, valleys, specific geologic structures)

- demographics

  - latest census results for area of concern

  - description, distance, and direction to nearby population centers

  - description, distance, and direction to nearby public facilities (e.g., schools, hospitals, parks)

  - description, distance, and direction to nearby industrial areas or facilities that may present potential hazards (including other nearby nuclear facilities)

- uses of land within the licensed facility or its proposed boundaries (i.e., residential, industrial, commercial, agricultural)

- description of nearby bodies of water and their uses

- meteorology

  - primary wind directions and average windspeeds

  - annual amount and forms of precipitation, as well as the design-basis values for accident analysis of maximum snow or ice load and probable maximum precipitation

  - type, frequency, and magnitude of severe weather (e.g., lightning, tornado, hurricane) and design-basis event summary descriptions for accident analysis

- hydrology

  - characteristics of nearby rivers, streams, and bodies of water as appropriate

  - depth to the water table and potentiometric surface map

  - ground water flow direction and velocity for the site

  - characteristics of the uppermost aquifer and any hydrogeological connections to other aquifers in the region

  - design-basis flood events used for accident analysis

- geology

  - onsite characteristics of soil types and bedrock

  - design-basis earthquake magnitudes and return periods used for accident analysis

  - description of other geologic hazards (e.g., mass wasting)

Review Interfaces

To ensure consistency, the listed SRP sections and documents interface with this section as follows:

- Review information about the facility and site and the different processes that will involve licensable material in Chapter 3 of the SRP.

- Review the facility and process descriptions in Chapter 8 of the SRP.

- Review information in the applicant's environmental report about the site geography, demographics, meteorology, hydrology, and geology.

### 1.3.4 Acceptance Criteria

#### 1.3.4.1 Regulatory Requirements

The regulations applicable to the areas of review in this SRP are 10 CFR 70.22 and 10 CFR 70.65(b)(1) and (2).

#### 1.3.4.2 Regulatory Guidance

No regulatory guides apply to site descriptions for a fuel cycle facility.

#### 1.3.4.3 Regulatory Acceptance Criteria

The reviewer should find the site description, including the site geography, demographics, meteorology, hydrology, and geology (see Section 1.3.3), acceptable if it meets the following acceptance criteria:

- The summary briefly describes site geography, including its location relative to prominent natural and manmade features (such as mountains, rivers, airports, population centers, schools, and commercial and manufacturing facilities). The summary also describes the site boundary and the controlled area.

- The summary provides population information on the basis of the most current available census data. To the extent possible, data reflect observations and measurements made over a period of years, especially for conditions that are expected to vary seasonally (e.g., precipitation, windspeed and direction, and ground water levels).

- The application addresses appropriate meteorological data, including a summary of design-basis values for accident analysis of maximum snow or ice load and probable maximum precipitation, as may be developed by the applicant and presented in the ISA Summary. The applicant presents appropriate design-basis values for lightning, high winds, tornado, hurricane, and other severe weather conditions that are applicable to the site.

- The application includes a summary description of the hydrology and geology (including seismicity) for the area and cites the design-basis flood event for which the facility may be safely shut down.

- The applicant's descriptions are consistent with the more detailed information presented in the ISA Summary, the environmental report, and the emergency plan, if applicable.

### 1.3.5 Review Procedures

#### 1.3.5.1 Acceptance Review

The staff will initially determine whether the application is complete and addresses all topics discussed in Section 1.3.3. The information in this section provides a general summary of the bases reported in the ISA Summary and is consistent with the applicant's environmental report and emergency plan. The applicant may include references to the more detailed data used to complete evaluations in the ISA Summary.

If the reviewer identifies significant deficiencies in the license application during the acceptance review, the staff should ask the applicant to submit additional material or revise the license application. This will ensure that the license application is technically sufficient before the start of the safety evaluation. Once the deficiencies in the license application have been addressed during the acceptance review, the NRC reviewer should examine the information on site descriptions and assess its acceptability by comparison with the acceptance criteria in Section 1.3.4.3.

For license renewals or license amendment applications, the details necessary to support the information in the site description summary may be referenced as a prior submittal or material included elsewhere in the application.

*1.3.5.2 Safety Evaluation*

The material in this section of the SRP is informational, as it summarizes material in the ISA Summary, environmental report, emergency plan, and other documents referenced by the applicant. No technical analysis is required, as the primary reference for the information is the ISA Summary. The applicant may also need to update this section to verify any information changes made in response to the staff's environmental, emergency management, and ISA Summary reviews.

## 1.3.6 Evaluation Findings

If the license application provides sufficient information and is consistent with the guidance in this SRP, the staff will conclude that this evaluation is complete and the applicant's site description is acceptable. The staff can document its review as follows:

> The staff has reviewed the site description for [name of facility] in accordance with Section 1.3 of the Standard Review Plan. [Name of facility] has adequately described and summarized general information pertaining to (1) the site geography, including its location relative to prominent natural and manmade features such as mountains, rivers, airports, population centers, schools, and commercial and manufacturing facilities, (2) population information using the most current available census data, (3) meteorology, hydrology, and geology for the site, and (4) applicable design-basis events. The review verified that the site description is consistent with the information used as a basis for the environmental report, emergency management plan, and ISA Summary.

## 1.4 References

*U.S. Code of Federal Regulations*, Chapter I, Title 10, "Energy," Part 70, "Domestic Licensing of Special Nuclear Material."

*U.S. Code of Federal Regulations*, Chapter I, Title 10, "Energy," Part 95, "Facility Security Clearance and Safeguarding of National Security Information and Restricted Data."

# 2. ORGANIZATION AND ADMINISTRATION

## 2.1 Purpose of Review

The purpose of the review of the applicant's organization and administration is to ensure that the proposed management hierarchy and policies will provide reasonable assurance that the licensee plans, implements, and controls site activities in a manner that ensures the safety of workers, the public, and the environment. The review also ensures that the applicant has identified and provided adequate qualification descriptions for key management positions.

## 2.2 Responsibility for Review

Primary:      Licensing Project Manager

Secondary:    None

Supporting:   Primary Reviewers for Other Standard Review Plan (SRP) Chapters
              (e.g., technical area chapters and management measures chapters)
              Fuel Facility Inspection Staff

## 2.3 Areas of Review

The organizational structure and associated administrative program proposed by the applicant should include administrative policies, procedures and management policies, and qualifications of key management positions and describe how these will provide reasonable assurance that the health, safety, and environmental (HS&E) protection functions will be effective.

For new facilities, or currently licensed facilities undergoing major modifications, the applicant should describe the comprehensive management policies and procedures that will be used to manage and closely monitor the facility design, engineering, construction, and modifications.

The application should address how the management policies ensure the establishment and maintenance of design and operations. The administrative and management policies should describe the relationships among major facility safety functions and programs, such as the integrated safety analysis, management measures for items relied on for safety (IROFS), radiation safety, nuclear criticality safety, fire safety, chemical safety, environmental monitoring, and emergency planning. The applicant should also describe its qualification criteria with regard to education (i.e., degree and field), training, and experience for key management positions. Management positions for which such criteria should be described include the facility manager, operations manager, shift supervisor, and managers for various safety and environmental disciplines. Alternative named management positions could be proposed. Qualification criteria should be described generally, in terms of academic credentials, formal continuing education, and work experience. For example, "bachelor's degree in nuclear engineering or related scientific or engineering field, with 5 years of experience managing the operations of a nuclear fuel manufacturing facility."

Review Interfaces

None

## 2.4 Acceptance Criteria

### 2.4.1 Regulatory Requirements

The regulations in Title 10 of the *Code of Federal Regulations* (10 CFR) 70.22(a)(6), 70.23(a)(2), and 70.62(d) require a management system and administrative procedures for the effective implementation of HS&E functions concerning the applicant's corporate organization, qualifications of the staff, and adequacy of the proposed equipment, facilities, and procedures to provide adequate safety for workers, the public, and the environment.

### 2.4.2 Regulatory Guidance

There are no regulatory guides specific to the organization and administration description of fuel cycle facilities.

### 2.4.3 Regulatory Acceptance Criteria

The application is acceptable if it meets the criteria identified below. The applicant's safety program description should include appropriate commitments relevant to these criteria.

The following criteria apply to new facilities or facilities undergoing major modifications (in addition to the criteria listed below for existing facilities):

- The applicant has identified and functionally described the specific organizational groups that are responsible for managing the design, construction, operations, and modifications of the facility or licensed activities. The application also includes organizational charts.

- Clear, unambiguous management controls and communications exist among the organizational units responsible for managing the design, construction, operations, and modifications of the facility or licensed activities.

- The personnel responsible for managing the design, construction, operation, and modifications of the facility or licensed activities have substantive breadth and level of experience and are appropriately available. The qualifications, responsibilities, and authorities for key supervisory and management positions with HS&E responsibilities are clearly defined in position descriptions that are accessible to all affected personnel and to the U.S. Nuclear Regulatory Commission (NRC), upon request.

- The applicant has described specific plans to commission the facility's startup and operation, including the transition from the startup phase to operations, under the direct supervision of the applicant's personnel responsible for safe operations. The application clearly describes the roles and responsibilities of the different functions engaged in these commissioning activities.

The following criteria apply to existing facilities:

- The applicant has identified and functionally described the specific organizational groups responsible for operating the facility and managing the development of design changes to the facility. The application also includes organizational charts.

- The qualifications, responsibilities, and authorities of key supervisory and management positions with HS&E responsibilities are clearly defined in position descriptions that are accessible to affected persons and to the NRC, upon request.

- In the organizational hierarchy, the HS&E organization(s) is independent of the operations organization(s), allowing it to provide objective HS&E audit, review, or control activities. "Independent" means that neither organization reports to the other in an administrative sense. (However, both may report to a common manager.) Lines of responsibility and authority are clearly drawn.

- The individual delegated overall responsibility for the HS&E functions has the authority to shut down operations if they appear to be unsafe and, in that case, must approve restart of shutdown operations or licensed activities.

- The activities essential for effective implementation of the HS&E functions are documented in formally approved, written procedures, prepared in compliance with a formal document control program.

- The applicant should commit to a simple mechanism, available for use by any person in the plant, for reporting potentially unsafe conditions or activities to the HS&E organization. Reported concerns should be promptly investigated, assessed, and resolved.

- The application clearly defines effective lines of communication and authority among the organizational units involved in the engineering, HS&E, and operations functions of the facility.

- The applicant has committed to establishing formal management measures required to ensure the availability and reliability of IROFS. Chapter 11 of this SRP discusses management measures.

- Written agreements exist with offsite emergency resources such as fire, police, ambulance and rescue units, and medical services. Chapters 7 and 8 of this SRP address these agreements in more detail.

The applicant's safety program description includes commitments relevant to meeting the acceptance criteria described above.

## 2.5 Review Procedures

### 2.5.1 Acceptance Review

The primary reviewer should evaluate the application to determine whether it addresses the areas of review discussed in Section 2.3. If the reviewer identifies significant deficiencies in the license application during the acceptance review, the staff should ask the applicant to submit additional material or revise the license application. This will ensure that the license application is technically sufficient before the start of the safety evaluation. Once the deficiencies in the license application have been addressed during the acceptance review, the NRC reviewer should examine the organization and administration information and assess its acceptability by comparison with the acceptance criteria in Section 2.4.3.

### 2.5.2 Safety Evaluation

The primary reviewer should perform a safety evaluation with respect to the acceptance criteria described in Section 2.4. The objective of the review is to ensure that the corporate-level management and technical support structure, as demonstrated by organizational charts and descriptions of functions and responsibilities, is clear with respect to assignments of primary responsibility. If necessary, the primary reviewer consults with the NRC inspection staff to verify that the applicant's management positions are adequately defined in terms of both numbers of persons and their responsibilities, authorities, and required qualifications. If necessary, the reviewer may visit the site to discuss and verify implementation of the acceptance criteria with facility management.

On the basis of the foregoing, the supporting staff reviewers determine the overall acceptability of the applicant's management system, management qualifications, organizational structure, and administrative procedures. The reviewers should determine whether the application satisfies the acceptance criteria of Section 2.4 and then prepare a safety evaluation report (SER) in accordance with Section 2.6.

## 2.6 Evaluation Findings

The staff's evaluation should verify that the license application provides sufficient information to satisfy the regulatory requirements of Section 2.4.1 and that the regulatory acceptance criteria in Section 2.4.3 have been appropriately considered in satisfying the requirements. On the basis of this information, the staff should conclude that this evaluation is complete. The reviewer should write material suitable for inclusion in the SER prepared for the entire application. The SER should include a summary statement of what was evaluated and the basis for the reviewers' conclusions. The staff can document the evaluation as follows:

> The staff has reviewed the organization and administration for [name of facility] according to Chapter 2 of the Standard Review Plan.
>
> (a)  For new facilities
>
> [Name of facility] described: (1) clear responsibilities and associated resources for the design, construction, operations, and modifications of the facility, and (2) its plans for managing the project. [Insert a summary statement of what was evaluated and why the reviewer finds the submittal acceptable.] The staff has reviewed these plans and commitments and concludes that they provide reasonable assurance that an acceptable organization, administrative policies, and sufficient qualified resources have been established or are committed to satisfy [name of facility]'s commitments for the design, construction, operations, and modifications of the facility.

(b)     For operating and new facilities

[Name of facility] described its organization and management policies for providing adequate safety management and management measures for the safe operation of the facility. [Insert a summary statement of what was evaluated and why the reviewer finds the submittal acceptable.] The staff has reviewed this information and concluded that [name of facility] has an acceptable organization, administrative policies, and sufficient qualified resources to provide for the safe operation of the facility under both normal and abnormal conditions.

## 2.7 **References**

*U.S. Code of Federal Regulations*, Chapter I, Title 10, "Energy," Part 70, "Domestic Licensing of Special Nuclear Material."

U.S. Nuclear Regulatory Commission, "Proposed Methods for Regulating Major Materials Licensees," NUREG-1324, Sections 3.1, "Organization Plan," and 3.2, "Managerial Controls and Oversight," February 1992.

# 3. INTEGRATED SAFETY ANALYSIS
# AND INTEGRATED SAFETY ANALYSIS SUMMARY

## 3.1 Purpose of Review

An integrated safety analysis (ISA) identifies potential accident sequences in the facility's operations, designates items relied on for safety (IROFS) to either prevent such accidents or mitigate their consequences to an acceptable level, and describes management measures to provide reasonable assurance of the availability and reliability of IROFS. Applicants for new licenses and persons holding licenses under Title 10 of the *Code of Federal Regulations* (10 CFR) Part 70, "Domestic Licensing of Special Nuclear Material," on September 18, 2000, must perform an ISA and submit a summary (referred to as an "ISA Summary") to the U.S. Nuclear Regulatory Commission (NRC) for approval. The ISA Summary focuses on higher risk accident sequences with consequences that could exceed the criteria of 10 CFR 70.61, "Performance Requirements." The ISA Summary is a synopsis of the results of the ISA and contains information specified in 10 CFR 70.65(b).

The ISA and supporting documentation (such as piping and instrumentation diagrams, criticality safety analyses, dose calculations, process safety information, and ISA worksheets) would be maintained on site at an existing facility. For an applicant seeking a license before commencing construction of a facility, full details concerning hardware, procedures, and programs usually would not exist. However, at the time of the operational readiness review[1] for a new facility, or major modifications to an existing facility, such details must exist to comply with the safety program requirements of 10 CFR Part 70, Subpart H, "Additional Requirements for Certain Licensees Authorized to Possess a Critical Mass of Special Nuclear Material." The level of detail that is acceptable in a license application and ISA Summary does not differ between existing and new facilities.

The NRC determines the acceptability of the applicant's ISA by reviewing a portion of the ISA documentation and any supporting documentation maintained on site and by reviewing and approving the applicant's ISA Summary which, although not part of the license application, is placed on the public docket. Neither the ISA nor the ISA Summary is incorporated as part of the license.

Reviewers must confirm that an ISA Summary meets the regulatory requirements of 10 CFR 70.65, "Additional Content of Applications," and, specifically, that suitable IROFS and management measures have been designated for high-risk accident sequences and that programmatic commitments to maintain the ISA and ISA Summary are acceptable. The term "programmatic" is used here to refer to the organization, criteria, methods, and practices for conducting activities important to safety, such as the ISA program, criticality and other safety discipline programs, and the management measures programs addressed in Chapter 11 of this Standard Review Plan (SRP). In fact, the primary purpose of the review conducted under the guidance of this chapter is to attain reasonable assurance that the applicant has established an ISA *program* that is, and will continue to be, in compliance with 10 CFR Part 70, Subpart H. Other chapters of this SRP offer guidance for review of management measures and other safety

---

[1] The operational readiness review is an assessment review inspection performed by a multidisciplinary inspection team to ensure that a plutonium or enrichment facility has been completed in accordance with the application or license, and so can be operated safely within the intended safety basis. For new facilities other than plutonium or enrichment facilities regulated under 10 CFR Part 70, Subpart H, or major modifications to existing ones, such reviews, though not strictly required, are normally conducted.

programs. This reasonable assurance of ISA program compliance is attained in part by a selective review of some of the ISA results, as described in Section 3.5 of this chapter under the subjects of "horizontal slice" and "vertical slice" reviews. However, it is not normally necessary to review the full details of all processes and IROFS in order to attain such reasonable assurance. An applicant may submit, for NRC approval, one ISA Summary for the entire facility, or multiple ISA Summaries for individual processes (or groups of processes) in the facility as they are completed. Reviewers of ISA Summaries for new and existing facilities may find it useful to examine the ISA and its supporting documentation to confirm the underpinnings of calculations, conclusions, and components of safety programs.

This chapter provides guidance for the NRC's review of two types of information submitted by applicants:

(1)     commitments regarding the applicant's safety program including the ISA, pursuant to the requirements of 10 CFR 70.62, "Safety Program and Integrated Safety Analysis"

(2)     ISA summaries submitted in accordance with 10 CFR 70.62(c)(3)(ii) and 10 CFR 70.65

In the case of license applications (either initial or renewal), applicants would submit both types of information. In the case of a license amendment, an applicant may submit either or both types of information, as needed, to address the areas amended.

The purpose of the review of the ISA Summary is to establish reasonable assurance that the applicant has performed the following tasks:

- Conducted an ISA of appropriate detail for each applicable process, using methods and staff adequate to achieve the requirements of 10 CFR 70.62(c)(1) and (2).

- Identified and evaluated, in the ISA, all credible events (accident sequences) involving process deviations or other events internal to the facility (e.g., explosions, spills, and fires) and credible external events that could result in facility-induced consequences to workers, the public, or the environment, that could exceed the performance requirements of 10 CFR 70.61. As a minimum, external events normally include the following:

    - natural phenomena such as floods, high winds, tornadoes, and earthquakes
    - fires external to the facility
    - transportation accidents and accidents at nearby industrial facilities

- Designated engineered and administrative IROFS and correctly evaluated the set of IROFS addressing each accident sequence, as providing reasonable assurance, through preventive or mitigative measures and through application of supporting management measures (discussed in Chapter 11 of this SRP) that the performance requirements of 10 CFR 70.61 are met.

## 3.2 Responsibility for Review

Primary:         Assigned Licensing Reviewer

Secondary:    Technical Specialists in Specific Areas

Supporting:   Fuel Facility Inspectors

## 3.3 Areas of Review

This chapter addresses two types of submittals: (1) those containing descriptive commitments regarding the safety program, including the ISA; and (2) ISA summaries. The descriptive commitments for the safety program should be found in license applications, renewals, and amendments. ISA summaries may be submitted for an entire existing facility, a new facility, a new process, or altered processes requiring revision of the ISA.

The safety program and ISA commitments and descriptions to be reviewed consist of (1) process safety information (10 CFR 70.62(b)), (2) methods used to perform the ISA, (3) qualifications of the team performing the ISA (10 CFR 70.62(c)(2)), (4) methods of documenting and implementing the results of the ISA, (5) procedures to maintain the ISA current when changes are made to the facility, and (6) management measures (10 CFR 70.62(d)). An ISA chapter in the license application will usually contain appropriate documentation of these commitments and descriptions. However, pursuant to Chapter 11 of this SRP, a separate chapter of an application may address the commitments to and descriptions of management measures.

An ISA Summary presents the results of ISA analyses performed for compliance with Subpart H of 10 CFR Part 70. This ISA Summary may be submitted with an application for a new license, a license renewal, or a license amendment but is not to be incorporated as part of the license.

The staff will review the ISA Summary submitted to the NRC and the portions of the ISA and ISA documentation maintained on site to determine the adequacy of the applicant's ISA. The contents of the ISA Summary, specified in 10 CFR 70.65, include the following nine topics:

(1)     general description of the site
(2)     general description of the facility
(3)     description of facility processes, hazards, and types of accident sequences
(4)     demonstration of compliance with 10 CFR 70.61 performance requirements
(5)     description of the ISA team qualifications and ISA methods
(6)     descriptive list of IROFS
(7)     description of acute chemical exposure standards used
(8)     descriptive list of sole IROFS
(9)     definition of the terms "credible," "unlikely," and "highly unlikely"

The documentation supporting the ISA (e.g., piping and instrumentation drawings, engineered IROFS boundary descriptions, criticality safety analyses, dose calculations, process hazards analysis, process safety information, ISA worksheets) will normally be maintained at the facility site. The reviewer may find it efficient to consult the ISA supporting documentation at the facility site to establish the completeness and acceptability of the ISA or, in the case of an existing facility, to visit the site to fully understand a process operation. For example, the reviewer could confirm that accident sequences that were not reported in the ISA Summary because they were not credible were correctly identified and analyzed in the ISA.

### 3.3.1 Safety Program and Integrated Safety Analysis Commitments

The NRC reviews the application to determine whether the applicant's commitments to establish a safety program and to perform and maintain an ISA are adequate. In the following, the phrase "process node" or "process" refers to a single, reasonably compact piece of equipment or workstation where a single unit process or processing step is conducted. A typical fuel cycle facility is divided into several major process lines or areas, each consisting of many process nodes. The areas of review for ISA commitments are as follows:

- The applicant's description of, and commitments to, a method for maintaining a current and accurate set of process safety information, including information on the hazardous materials, technology, and equipment used in each process. The applicant should explain this activity in detail in the description of its configuration management program (Section 11.1).

- The applicant's description of, and commitments to, requirements for ISA team training and qualifications (Section 11.4) for those individuals who will conduct and maintain the ISA and ISA Summary.

- The applicant's description of, and commitments to, ISA methods, method selection criteria, or specific methods to be used for particular classes of process nodes (usually process workstations). The review of the ISA method includes evaluating the applicant's methods in the following specific areas:

    - hazard identification
    - process hazard analysis (accident identification)
    - accident sequence construction and evaluation
    - consequence determination and comparability to 10 CFR 70.61
    - likelihood categorization for determining compliance with 10 CFR 70.61

- The applicant's description of, and commitments to, management procedures for conducting and maintaining the ISA. Specific review areas include the following applicant procedures:

    - performance of, and updates to, the ISA
    - review responsibility
    - ISA documentation
    - reporting of ISA Summary changes per 10 CFR 70.72(d)(1) and (3)
    - maintenance of ISA records per 10 CFR 70.62(a)(2)

### 3.3.2 Integrated Safety Analysis Summary and Documentation

The NRC reviews the ISA Summary and, if necessary, the ISA and supporting ISA documentation to determine whether there is reasonable assurance that the applicant has performed a systematic evaluation of the hazards and has identified credible accident sequences, IROFS, and management measures that satisfy the performance requirements of 10 CFR 70.61. The NRC confirms that credible accidents that result in a release of radioactive material, a nuclear criticality event, or any other exposure to radiation resulting from use of licensed material that exceeds the exposure limits stated in 10 CFR 70.61 are "highly unlikely" or "unlikely," as appropriate. In addition, the NRC reviews accidents involving hazardous chemicals produced from licensed materials. Hazardous chemical include chemicals that are licensed materials or have licensed materials as precursor compounds, or substances that physically or chemically interact with licensed materials and that are toxic, explosive, flammable, corrosive, or reactive to the extent that they endanger life or health. These include substances that are commingled with licensed material or are produced by a reaction with licensed material. If a chemical accident has the potential to cause, or reduce protection from, a radiation exposure accident, then it also must be addressed (see Chapter 6 for more information on chemical process safety). On the other hand, accident sequences having unmitigated consequences that will not exceed the performance requirements of 10 CFR 70.61(c), once identified as such, do not require reporting in the ISA Summary.

The areas of review for the ISA Summary are as follows:

- Site: The site description in the ISA Summary (see Section 1.3) focuses on those factors that could affect safety, such as geography, meteorology (e.g., high winds and flood potential), seismology, demography, and nearby industrial facilities and transportation routes.

- Facility: The facility description in the ISA Summary focuses on features that could affect potential accidents and their consequences. Examples of these features are facility location, facility design information, and the location and arrangement of buildings on the facility site.

- Processes, Hazards, and Accident Sequences: The process description in the ISA Summary addresses each process that was analyzed as part of the ISA. Specific areas reviewed include basic process function and theory, functions of major components and their operation, process design and equipment, and process operating ranges and limits. This description must also include a list of the hazards (and interactions of hazards) for each process and the accident sequences that could result from such hazards and for which unmitigated consequences could exceed the performance requirements of 10 CFR 70.61.

- Demonstration of Compliance with 10 CFR 70.61: For each applicable process, this section presents the following information that should be developed in the ISA to demonstrate compliance with the performance criteria of 10 CFR 70.61:

  – postulated consequences and comparison to the consequence levels identified in 10 CFR 70.61, as well as information, such as inventory and release path factors supporting the results of the consequence evaluation

  – information showing how the applicant established the likelihoods of accident sequences that could exceed the performance requirements of 10 CFR 70.61

  – information describing how designated IROFS protect against accident sequences that could exceed the performance requirements of 10 CFR 70.61

- information on management measures applied to the IROFS (addressed in greater detail in Chapter 11)

- information on how the criticality monitoring requirements of 10 CFR 70.24, "Criticality Accident Requirements," are met

- if applicable, ways that the baseline design criteria of 10 CFR 70.64, "Requirements for New Facilities or New Processes at Existing Facilities," are addressed

- **Team Qualifications and ISA Methods:** This section should discuss the applicant's ISA team qualifications and ISA methods, as described in the ISA Summary. (If methods are adequately described in the license application, the applicant will not need to duplicate this information in the ISA Summary. The ISA Summary should include specific examples of the application of ISA methods to enable the reviewer to understand their selection and use.)

- **List of IROFS:** This list describes the IROFS for all intermediate- and high-consequence accidents in sufficient detail to permit an understanding of their safety function.

- **Chemical Consequence Standards:** This discussion identifies the applicant's quantitative standards for assessing the chemical consequence levels specified in 10 CFR 70.61, as described in the ISA Summary.

- **List of Sole IROFS:** This list identifies those IROFS that are the sole item preventing or mitigating an accident for which the consequences could exceed the performance requirements of 10 CFR 70.61.

- **Definitions of "Unlikely," "Highly Unlikely," and "Credible":** The applicant must define the terms "unlikely," "highly unlikely," and "credible," as used in the ISA Summary.

The regulations in 10 CFR 70.65(b) list the types of information required to be submitted in an ISA Summary. This includes generic information, such as site description, ISA methods, and ISA team qualifications. This also includes process-specific information, such as a list of IROFS, general descriptions of types of accident sequences, and "information demonstrating compliance with 10 CFR 70.61." To meet the latter requirement, an applicant would have to provide, at a minimum, likelihood and consequence information for each type of process accident sequence identified in the ISA Summary. To evaluate the effectiveness of the applicant's likelihood and consequence evaluation methods, the reviewer should also examine the analyses of some accident sequences that are not reported in the ISA Summary for which the applicant established that consequences will not exceed the performance requirements of 10 CFR 70.61.

In some simple cases, the information normally contained in the ISA Summary process descriptions and list of IROFS might be sufficient to enable the reviewer to understand how compliance is achieved when considered with the description of ISA likelihood evaluation methods and criteria. However, in general, the applicant should describe how its ISA team evaluated a credible accident likelihood to be "highly unlikely" or "unlikely."

The reviewer should evaluate the efficacy of the applicant's ISA methods. To do this, in addition to reviewing the description of the ISA methods, the reviewer will need to understand how these methods have been applied in practice to the wide diversity of process safety designs in the facility. Examples in the ISA Summary of how the methods are applied to a representative sample of processes would help the reviewer to understand the applicant's ISA methods. However, if the ISA Summary does not include examples providing details of how the methods were applied, such information may be available at the applicant's site, as part of the overall safety information records.

The NRC review of the applicant's example accident sequence evaluations included in the ISA Summary is not a substitute for the "vertical slice" and "horizontal" reviews that should be performed using detailed information at the site. The NRC must select this onsite evaluation of ISA documentation and processes to confirm that the ISA was actually performed as described in the ISA Summary.

## 3.4 Acceptance Criteria

### 3.4.1 Regulatory Requirements

The regulation in 10 CFR 70.62 specifies the requirement to establish and maintain a safety program, including performance of an ISA. Paragraph (c) of 10 CFR 70.62 specifies requirements for conducting an ISA, which include a demonstration that credible high-consequence and intermediate-consequence events meet the safety performance requirements of 10 CFR 70.61. The requirement to prepare and submit an ISA Summary for NRC approval, stated in 10 CFR 70.65(b), also describes the contents of an ISA Summary. The regulation in 10 CFR 70.72, "Facility Changes and Change Process," set forth requirements for maintaining a current ISA and other safety program documentation when changes are made to the site, structures, processes, systems, equipment, components, computer programs, and activities of personnel; however, the ISA Summary need be updated only annually.

The information to be included in the ISA Summary can be divided into four categories: (1) site and facility characteristics, (2) ISA methods, (3) hazards and accident analysis, and (4) IROFS. Table 3.1 summarizes the information requirements of each category, the corresponding regulatory citation, and the section of this chapter that describes the expectations for such information.

## Table 3.1 Information Requirements for the ISA Summary

| Information Category and Requirement | 10 CFR Part 70 Regulatory Citation | NUREG-1520, Chapter 3 Section Reference |
|---|---|---|
| **_Site and Facility Characteristics_:**<br>Site description | 70.65(b)(1) | 3.4.3.2(1) |
| Facility description | 70.65(b)(2) | 3.4.3.2(2) |
| Criticality monitoring and alarms | 70.65(b)(4) | 3.4.3.2(4c) |
| Compliance with baseline design criteria, criticality monitoring, and alarms | 70.64 (if applicable)<br>70.65(b)(4) | 3.4.3.2(4d) |
| **_ISA Methods_** | | |
| ISA method(s) description | 70.65(b)(5) | 3.4.3.2(5) |
| ISA team description | 70.65(b)(5) | 3.4.3.2(5) |
| Quantitative standards for acute chemical exposures | 70.65(b)(7) | 3.4.3.2(7) |
| Definition of "unlikely," "highly unlikely," and "credible" | 70.65(b)(9) | 3.4.3.2(9) |
| **_Hazards and Accident Analysis_** | | |
| Description of processes analyzed | 70.65(b)(3) | 3.4.3.2(3a) |
| Identification of hazards | 70.65(b)(3) | 3.4.3.2(3b) |
| Description of accident sequences | 70.65(b)(3) | 3.4.3.2(3c) |
| Characterization of high- and intermediate-consequence accident sequences | 70.65(b)(3) | 3.4.3.2(3c) |
| **_Items Relied on for Safety_** | | |
| List and description of IROFS | 70.65(b)(6) | 3.4.3.2(6) |
| Description of IROFS' link to accident sequences to show 10 CFR 70.61 compliance | 70.65(b)(6) | 3.4.3.2(4) and (6) |
| IROFS management measures | 70.65(b)(4) | 3.4.3.2(4b) and (6) |
| List of sole IROFS | 70.65(b)(8) | 3.4.3.2(8) |

## 3.4.2 Regulatory Guidance

NUREG-1513, "Integrated Safety Analysis Guidance Document," issued May 2001, contains guidance applicable to performing an ISA and documenting the results. NUREG/CR-6410, "Nuclear Fuel Cycle Facility Accident Analysis Handbook," issued March 1998, provides guidance on acceptable methods for evaluating the chemical and radiological consequences of potential accidents. NUREG-1601, "Chemical Process Safety at Fuel Cycle Facilities," issued August 1997, provides guidance on chemical safety practices acceptable for compliance with the regulations.

### 3.4.3 Regulatory Acceptance Criteria

The acceptance criteria for an ISA are derived from and support compliance with the relevant requirements of 10 CFR Part 70. The ISA will form the basis for the safety program by identifying potential accidents, designating IROFS and management measures, and evaluating the likelihood and consequences of each accident sequence for compliance with the performance requirements of 10 CFR 70.61. Some of the acceptance criteria address the programmatic commitments made by the applicant to perform and maintain an ISA. The remainder of the criteria address the ISA results, as documented in the ISA Summary, and whether those documented results demonstrate that the applicant's IROFS and management measures can reasonably be expected to ensure that the relevant accident sequences will meet the performance requirements of 10 CFR 70.61. The acceptance criteria are thus intended to support the ultimate finding of the license review that, based on the information submitted and reviewed, there is reasonable assurance that the proposed facility, IROFS, safety programs, and management measures conforming to the commitments in the application comply with the regulations and provide adequate protection of public health and safety.

A high level of detail describing the process designs and IROFS might not be submitted with the license application or ISA Summary. In other words, the applicant might not provide information about all the components in a system, because not every component would be a safety-related component. In particular, for proposed new facilities, the level of detail may be limited since the hardware has not actually been fabricated. However, the applicant must describe the IROFS in enough detail to permit an understanding of the intended safety function and to permit an assessment that it is capable of the reliability expected of it in the evaluation of likelihoods of accident sequences. The NRC staff may obtain additional details for processes selected for the vertical slice review by visiting the applicant's site. While there may be an *actual* difference in the level of detail known about processes and IROFS, as documented at the applicant's site, for existing and proposed new facilities, the minimum level of detail that is sufficient in descriptions of processes and IROFS, as documented in the ISA Summary, does not differ between existing and proposed new facilities.

The purpose of the review, and its acceptance criteria, for most facilities, is primarily to permit a finding that the applicant's safety program, including the ISA program as described, provides reasonable assurance that compliance will be achieved. However, to generate the ISA Summary, which is a required submission, the applicant must first perform an ISA. This in turn requires that the applicant identify process designs, accident sequences, and IROFS. These latter items are not programmatic, but are elements of design and analysis of design. Attainment of reasonable assurance that the ISA program is and will be effective does not usually require that all safety elements and IROFS be reviewed in full detail, nor is it required that the applicant's description of IROFS and process designs be at the level of detail that will eventually exist at the time of operations (see the discussion of vertical slice review in Section 3.5). The requisite level of detail to achieve reasonable assurance may vary among processes, depending on factors such as use of established technology, commitment to standards, applicant expertise, industry experience, safety margins, and inherent difficulty in achieving the safety function. However, the underlying requirements for the descriptions are exactly the same for each process and IROFS; namely, "...a description of each process...in sufficient detail to understand the theory of operation..." (10 CFR 70.65(3)); and "a description of IROFS...in sufficient detail to understand their functions in relation to the performance requirements..." (10 CFR 70.65(8)). Thus, the requirements for new technology are no different than those for old technology, but more explanatory detail may be necessary to meet the requirements related to "sufficient detail to understand."

### 3.4.3.1 Safety Program and Integrated Safety Analysis Commitments

This section discusses the acceptance criteria for license commitments pertaining to the facility's safety program including the performance of an ISA. A number of specific safety program requirements related to the ISA appear in 10 CFR Part 70. Section 3.4.3.2 presents the acceptance criteria for the content of the ISA Summary. These include the primary requirements that an ISA be conducted and that, based on the ISA Summary submitted, there is reasonable assurance that the applicant's facility and safety program complies with the ISA requirements of 10 CFR Part 70, Subpart H, including the performance requirements of 10 CFR 70.61. For each component of the safety program, several elements may be necessary, including organization, assignment of responsibilities, management policies, required activities, written procedures for activities, use of industry consensus standards, and technical safety practices, among others.

Procedures and industry standards for hardware safety controls vary according to the type of equipment and by the degree of reliability and performance required in specific applications. For this reason, blanket commitments to apply all standards in all cases may not appear in the license application. However, some standards for engineering practices and hardware and software design or analysis are generic. Hence, an applicant may specify a general commitment to such a generic standard or may make conditional commitments to standards, subject to specified applicability criteria. The purpose of such commitments is to support likelihood or other performance evaluations for compliance with the regulations. NRC guidance has endorsed some standards, possibly with exceptions. Such commitments to standards are acceptable if they are consistent with their use in demonstrating compliance and with specific NRC guidance.

Among those engineering practices and standards that are generically applicable to IROFS and safety controls are those that apply to personnel activities relevant to administrative controls, management measures, or human-machine interfaces. This area is called human factors engineering. Human factors engineering should generally be part of the safety program. Human factors practices should be incorporated into the applicant's safety program sufficiently to ensure that IROFS and management measures perform their functions in meeting the requirements of 10 CFR Part 70. Appendix E to this chapter describes areas of review and acceptance criteria for human factors engineering in the context of 10 CFR Part 70 for fuel cycle facilities.

The applicant's commitments for each of the three elements of the safety program defined in 10 CFR 70.62(a) should be acceptable if the applicant does the following:

(1) Process Safety Information

    a.    The applicant commits to compiling and maintaining an up-to-date database of process safety information. Written process safety information will be used in updating the ISA and in identifying and understanding the hazards associated with the processes. The compilation of written process safety information should include information pertaining to the following:

i.    The description of hazards of all materials used or produced in the process, which should include information on chemical and physical properties (such as toxicity, acute exposure limits, reactivity, and chemical and thermal stability) such as are included on Material Safety Data Sheets (meeting the requirements of 29 CFR 1910.1200(g)).

ii.   The discussion of the technology of the process should include a block flow diagram or simplified process flow diagram, a brief outline of the process chemistry, safe upper and lower limits for controlled parameters (e.g., temperature, pressure, flow, and concentration), and evaluation of the health and safety consequences of process deviations.

iii.  The description of the equipment used in the process should include general information on topics such as the materials of construction, piping and instrumentation diagrams, ventilation, design codes and standards employed, material and energy balances, IROFS (e.g., interlocks, detection, or suppression systems), electrical classification, and relief system design and design basis.

b.    The applicant includes procedures and criteria for changing the ISA, along with a commitment to design and implement a facility change mechanism that meets the requirements of 10 CFR 70.72. The applicant should discuss the evaluation of the change within the ISA framework, as well as procedures and responsibilities for updating the facility's ISA.

c.    The applicant commits to engage personnel with appropriate experience and expertise in engineering and process operations to maintain the ISA. The ISA team for a process should consist of individuals who are knowledgeable in the facility's ISA methods and the operation, hazards, and safety design criteria of the particular process.

(2)   ISA

a.    The applicant conducts and commits to maintaining an ISA of appropriate complexity for each process, such that it identifies (i) radiological hazards, (ii) chemical hazards that could increase radiological risk, (iii) facility hazards that could increase radiological risk, (iv) potential accident sequences, (v) consequences and likelihood of each accident sequence, and (vi) IROFS including the assumptions and conditions under which they support compliance with the performance requirements of 10 CFR 70.61. The application is acceptable if it describes sufficiently specific methods and criteria that would be effective in accomplishing each of these tasks. Such effective methods and criteria are described in NUREG-1513, NUREG/CR-6410, item (5) of Section 3.4.3.2 of this SRP, and Appendix A to this chapter.

b.    The applicant commits to keeping the ISA and its supporting documentation accurate and up to date by means of a suitable configuration management system and to submitting changes to the ISA Summary to the NRC, in accordance with 10 CFR 70.72(d)(1) and (3). The ISA must account for any changes made to the facility or its processes (e.g., changes to the site, operating procedures, or control systems). Management policies, organizational responsibilities, revision timeframe, and procedures to perform and approve revisions to the ISA should be outlined succinctly. The applicant commits to evaluating any facility changes or changes in the process safety information that may alter the parameters of an accident sequence by means of the facility's ISA methods. For any revisions to the ISA, the applicant commits to using personnel with qualifications similar to those of ISA team members who conducted the original ISA.

c. The applicant commits to training personnel in the facility's ISA methods and/or using suitably qualified personnel to update and maintain the ISA and ISA Summary.

d. The applicant commits to evaluating proposed changes to the facility or its operations by means of the ISA methods and to designating new or additional IROFS and appropriate management measures as required. The applicant also agrees to promptly evaluate the adequacy of existing IROFS and associated management measures and to make any required changes that may be affected by changes to the facility and/or its processes. If a proposed change results in a revised accident sequence in the ISA Summary or increases the consequences and/or likelihood of a previously analyzed accident sequence within the context of 10 CFR 70.61, the applicant commits to promptly evaluating the adequacy of existing IROFS and associated management measures and to making necessary changes, if required.

e. The applicant commits to addressing any unacceptable performance deficiencies in the IROFS that are identified through updates to the ISA.

f. The applicant commits to maintaining written procedures on site.

g. The applicant commits to establishing all IROFS (if not already established) and to maintaining them so that they are available and reliable when needed.

In citing industry consensus standards, the applicant should delineate specific commitments in the standards that will be adopted. The applicant should provide justification if it has not adopted all of the required elements of a standard.

(3) Management Measures

The applicant commits to establishing management measures (which are evaluated using SRP Chapter 11) that constitute the principal mechanism for ensuring the reliability and availability of each IROFS.

### 3.4.3.2 Integrated Safety Analysis Summary and Documentation

Information in the ISA Summary should provide the basis for the reviewer to conclude that there is reasonable assurance that the identified IROFS will satisfy the performance requirements of 10 CFR 70.61. To do this, the reviewer must conclude that the applicant's ISA program has the capability to identify appropriate IROFS and that IROFS identified in the ISA Summary are adequate to control the potential accidents of concern at the facility. The accidents of concern are those that would have consequences at the high and intermediate levels, absent any preventive or mitigative controls. In this context, adequacy means the capability of the IROFS to prevent the related accidents with sufficient reliability, or to sufficiently mitigate their consequences, so that the performance requirements of 10 CFR 70.61 can be met. To support such a review, the ISA Summary must include sufficient information about an accident sequence and the proposed IROFS to allow the reviewer to assess the contributions of the IROFS to prevention or mitigation. The ISA Summary must contain enough information concerning the ISA methods and the qualifications of the team that performed the ISA and any other resources employed to give the reviewer confidence that the list of potential accidents identified is reasonably complete.

In addition, the reviewer needs to determine that appropriate management measures will be in place to ensure the availability and reliability of the identified IROFS, when needed. Chapter 11 of this SRP addresses the review of designated management measures

The following acceptance criteria address each of the content elements of the ISA Summary required by 10 CFR 70.65(b). For new facilities, the reviewer should also evaluate those aspects of the design that address the baseline design criteria of 10 CFR 70.64 applicable to individual processes. Thus, the following nine content elements have defined acceptance criteria:

(1)     general description of the site
(2)     general description of the facility
(3)     description of facility processes, hazards, and types of accident sequences
(4)     demonstration of compliance with 10 CFR 70.61 performance requirements
(5)     description of the ISA team qualifications and ISA methods
(6)     descriptive list of IROFS
(7)     description of acute chemical exposure standards used
(8)     descriptive list of sole IROFS
(9)     definitions of "credible," "unlikely," and "highly unlikely"

Detailed acceptance criteria for each element of the ISA Summary follow:

(1)     Site. The description in the ISA Summary of the site for processing nuclear material is considered acceptable if the applicant includes, or references, the following safety-related information, with emphasis on those factors that could affect safety:

      a.     A description of the site geography, including its location, taking into account prominent natural and manmade features such as mountains, rivers, airports, population centers, possibly hazardous commercial and manufacturing facilities, transportation routes, etc., adequate to permit evaluation of (i) the likelihoods of accidents caused by external factors and (ii) the consequences of potential accidents.

      b.     Population information, based on the most recent census data, that shows population distribution as a function of distance from the facility, adequate to permit evaluation of regulatory requirements, including exposure of the public to consequences listed in 10 CFR 70.61.

c.   Characterization of natural phenomena (e.g., tornadoes, hurricanes, floods, and earthquakes) and other external events sufficient to allow assessment of their impact on facility safety and their likelihood of occurrence. At a minimum, the 100-year flood should be postulated, consistent with U.S. Army Corps of Engineers flood plain maps. The applicant should also provide earthquake accelerations for the site associated with a 250-year and 500-year earthquake. The discussion should identify all design-basis natural events for the facility, indicate which events are considered incredible, and describe the basis for that determination. The assessment should also indicate which events could occur without adversely impacting safety.

(2)   Facility. The description of the facility is considered acceptable if the applicant identifies and describes the general features that affect the reliability or availability of IROFS. If such information is available elsewhere in the application, reference to the appropriate sections is considered acceptable. The information provided should adequately support an overall understanding of the facility structure and its general arrangement. As a minimum, the applicant should adequately identify and describe the following:

a.   the facility location and the distance from the site boundary in all directions, including the distance to the nearest resident and distance to boundaries in the prevailing wind directions

b.   restricted area and controlled area boundaries

c.   design information regarding the resistance of the facility to failures caused by credible external events, when those failures may produce consequences exceeding those identified in 10 CFR 70.61

d.   the location and arrangement of buildings on the facility site

(3)   Processes, Hazards, and Accident Sequences

a.   Processes. The descriptions of processes in the ISA Summary must include all processes in which upset conditions could credibly lead to accidents with high or intermediate consequences. No areas or processes can be omitted, unless screened out because the accidents are non-credible. The description in the ISA Summary of the processes analyzed as part of the ISA (10 CFR 70.62(c)(1)(i–vi)) is considered acceptable if it describes the following features in sufficient detail to permit an understanding of the theory of operation and to assess compliance with the performance requirements of 10 CFR 70.61. A description at a systems level is acceptable, provided that it permits the NRC reviewer to adequately evaluate (1) the completeness of the hazard and accident identification tasks and (2) the likelihood and consequences of the accidents identified. If the information is available elsewhere in the application and is adequate to support the purposes of the ISA Summary, reference to the appropriate sections is considered acceptable. The descriptions of processes must permit an understanding of how the set of IROFS in that process could reliably perform their safety function for each high- and intermediate-consequence accident sequence. Hence, all process designs must be described in sufficient detail to reasonably permit identification of all accident sequences and IROFS to prevent or mitigate them.

The level of detail in process safety documentation held at the site would normally be greater than the descriptions in the ISA Summary and may include some or all of the information listed as items i through iv below, as needed.

i.     Basic process function and theory includes a general discussion of the basic theory of the process. Normally, this would include the following:

- parameters to be controlled and strategy for complying with 10 CFR 70.61

- chemical or mechanical theory principles, materials, and quantities needed to understand the hazards and safety functions

- normal and potential transport and changes in materials in the process

ii.     Major components include the general arrangement, function, and operation of major components in the process. If appropriate, it could also include arrangement drawings and process schematics showing the major components and instrumentation, and flowsheets showing compositions of the various process streams.

iii.     Process design and equipment include a discussion of process design, equipment, and instrumentation that is sufficiently detailed to permit an adequate understanding of the results of the ISA. As appropriate, it includes schematics indicating safety interrelationships of parts of the process. In particular, it is usually necessary for criticality safety to diagram the location and geometry of the fissile and other materials in the process, for both normal and bounding abnormal conditions. This can be done using either schematic drawings or textual descriptions indicating the location and geometry of fissile materials, moderators, etc., sufficient to permit an understanding of how the IROFS limit the mass, geometry, moderation, reflection, and other factors.

iv.     Process safety limits and margins on variables (e.g., temperatures, pressures, flows, fissile mass, enrichment, and composition) that are controlled by IROFS to ensure safe operations of the process, should be specified, because these limits and margins would be needed to understand the likelihood of failure assigned by the applicant to the IROFS. For example, if a process is designed, and an IROFS procedure specified, to ensure critical mass control by double-batching proof, the margin from a single batch to the subcritical limit should be specified. Traditionally, the single batch is 45 percent of the subcritical limit.

b.     Hazards. The description of process hazards provided in the ISA Summary is acceptable if it identifies, for each process, all types of hazards that are relevant to determining compliance with the performance criteria of 10 CFR 70.61. That is, the acceptance criterion is completeness. All hazards that could result in an accident sequence in which the consequences could exceed the performance requirements of 10 CFR 70.61 should be listed, even if later analysis of a particular hazard shows that resulting accident sequences do not exceed these limits. Otherwise, the reviewer cannot determine completeness. General exclusion from consideration of certain hazards for an entire facility can be justified by bounding case analyses showing that, for the conditions or credible inventories on site, the performance requirements of 10 CFR 70.61 cannot be exceeded. In this case, the bounding inventories or conditions, if under the control of the applicant, become IROFS.

Any locations where hazardous regulated material, including fissile material, could accidentally be located should also be considered. Improper screening out of locations and processes that are not normally hazardous, but that could become so in upset conditions, can lead to a failure to apply IROFS to prevent such upsets and potential accidents arising from them.

The list of process hazards is acceptable if the ISA Summary provides the following information:

i.      a list of materials (radioactive, fissile, flammable, and toxic) or conditions that could result in hazardous situations (e.g., loss of containment of licensed nuclear material), including the maximum intended inventory amounts and locations of the hazardous materials at the facility

ii.     potential interactions among materials or conditions that could result in hazardous situations

c.      Accident Sequences. The general description of types of accident sequences in the ISA Summary is acceptable if the reviewer can determine the following:

i.      The applicant has identified all types of accidents for which the consequences could exceed the performance requirements of 10 CFR 70.61. The level of detail required in describing accidents is closely related to the level of detail in describing IROFS, as many events leading to consequences of concern in 10 CFR 70.61 are failures of IROFS. It is not usually necessary to specify all modes and mechanisms by which the IROFS failure could occur in order to understand the role that the IROFS plays in preventing or mitigating the accident.

ii.     The applicant has identified how the IROFS listed in the ISA Summary protect against each such type of accident.

To satisfy the requirement that all accidents be identified, the applicant should describe the process design in sufficient detail. In particular, all IROFS need to be specified. The level of detail in specifying the process and IROFS should be sufficient to permit a reasonable understanding as to how the safety function will be performed so as to meet the performance requirements of 10 CFR 70.61.

General types of accident sequences differ if they consist of a different set of IROFS failures. Thus, several processes, each using a set of IROFS that is functionally of the same type (e.g., having the same mechanical, physical, and/or electrical principle of operation), can be summarized as a single type of accident sequence and listed only once. However, the individual processes covered by this system should be individually identified in a way that the reviewer can determine the application's completeness in addressing all processes.

For this reason, it is not generally acceptable as a description of an accident to merely list the type of hazard, or the controlled parameters, without referencing the items relied on to control the parameters or hazard. The description of general types of accident sequences is acceptable if it covers all types of sequences, initiating events, and IROFS failures. Initiating events can be (1) an external event such as a hurricane or earthquake, (2) a facility event external to the process being analyzed (e.g., fires, explosions, failures of other equipment, flooding from facility water sources), (3) deviations from normal operations of the process (credible abnormal events), or (4) failures of an IROFS in the process. Human errors that are initiating events would generally be administrative IROFS failures. The description of a general type of accident sequence is acceptable if it permits the reviewer to determine how each accident sequence for which the consequences could exceed the performance requirements of 10 CFR 70.61 is protected against by IROFS or a system of IROFS.

One acceptable way to do this is to show a fault tree on which the basic events are IROFS failures. Another acceptable method is to provide a table in which each row displays the events in an accident sequence, such as in Appendix A to this chapter, Table A-7, where, in general, each event is a failure of an IROFS. Another acceptable way is to provide a narrative summary for each process describing the sequence of events in each type of accident.

To demonstrate completeness, the process hazard analysis identifying general types of accident sequences must use systematic methods and consistent references. Therefore, each description of a general type of accident sequence is acceptable if it meets the following criteria:

i.      An acceptable method of hazard identification and process hazard analysis is used in accordance with the criteria of NUREG-1513.

ii.     The selected method is correctly applied.

iii.    The applicant does not overlook any type of accident sequence for which the consequences could exceed the performance requirements of 10 CFR 70.61. A key test of whether a type of accident has been overlooked is whether IROFS have been identified to meet the performance requirements.

iv.     The applicant uses a method of identifying facility processes that ensures identification of all processes.

During the early phases of an ISA, accidents will be identified for which the consequences may initially be unknown. These accidents will later be analyzed and may be shown to have consequences that are less than the levels identified in 10 CFR 70.61.

The ISA Summary need not list as a separate type of accident sequence, every conceivable permutation of an accident. Accidents having characteristics that all fall in the same categories can be grouped as a single type of accident in the ISA Summary provided that the following conditions are met:

i.      The initiating IROFS failures or events have the same effect on the system.
ii.     They all consist of failures of the same IROFS or system of IROFS.
iii.    They all result in violation of the safety limit on the same parameter.
iv.     They all result in the same type and severity categories of consequences.

(4)     Information Demonstrating Compliance with the Performance Requirements of 10 CFR 70.61

    a.  Accident Sequence Evaluation and IROFS Designation. The regulation in 10 CFR 70.65(b)(4) requires that the ISA Summary contain "information that demonstrates the licensee's compliance with the performance criteria of 10 CFR 70.61 . . ." Since the requirements of 10 CFR 70.61 are expressed in terms of consequences and likelihoods of events, the ISA Summary should provide sufficient information to demonstrate the following:

        i.      Credible high-consequence events are highly unlikely.

        ii.     Credible intermediate-consequence events are unlikely.

        iii.    Under normal and credible abnormal conditions, all nuclear processes are subcritical.

    The performance requirements of 10 CFR 70.61 have three elements, including completeness, consequences, and likelihood.

    "Completeness" refers to the requirement that the ISA address *each* credible event. "Consequence" refers to the magnitude of the chemical and radiological doses of the accident and is the basis for classification of an accident as a high- or intermediate-consequence event as described in 10 CFR 70.61. "Likelihood" refers to the requirement in 10 CFR 70.61 that intermediate-consequence events be "unlikely" and high-consequence events be "highly unlikely." Thus, the information provided must address each of these three elements.

    To be acceptable, the information provided must correspond to the ISA methods, consequence, and likelihood definitions described in the submittal. The information must also show the basis for and results of applying these methods to each process. In addition, the information must show that the methods have been properly applied in each case.

    The information showing completeness, consequences, and likelihood for accident sequences can be presented in various formats, including logic diagrams, fault trees, or tabular summaries. Appendix A to this chapter shows one example of how an application can present this information.

    Each of these performance requirements (completeness, consequences, and likelihood) is discussed below.

i.    Completeness is demonstrated by correctly applying an appropriate accident identification method, as described in NUREG-1513. Completeness can be effectively displayed by using an appropriate diagram or description of the identified accidents.

ii.    Consequence information in the ISA Summary is acceptable for showing compliance with 10 CFR 70.61 provided that the following conditions are met:

- For each accident for which the consequences could exceed the performance requirements of 10 CFR 70.61, the ISA Summary includes an estimate of its quantitative consequences (doses, chemical exposures, criticality) in a form that can be directly compared with the consequence levels in 10 CFR 70.61 or includes a reference to a value documented elsewhere in the ISA Summary that applies to or bounds that accident.

- The consequences were calculated using a method and data consistent with NUREG/CR-6410, or another method described and justified in the methods description section of the ISA Summary.

- All consequences that could result from the accident sequence have been evaluated. That is, if an accident can result in a range of consequences, all possibilities must be considered, including the maximum source term and most adverse weather that could occur. In other words, because of possible variations in weather or other conditions, the consequences of a type of potential accident may vary. If, for some such conditions, the consequences will be high, then the subset of such accidents resulting in high consequences are a "high consequence accident sequence" in the ISA, even though for average conditions, such high consequences would not result. If such conditions are *unlikely* to occur, credit can be taken for this in the evaluation of likelihood.

- The ISA Summary correctly assigns each type of accident to one of the consequence categories of 10 CFR 70.61 (namely, high or intermediate).

  Unshielded nuclear criticality accidents are considered to be high-consequence events, because the radiation exposure that an individual could receive exceeds the acute 1-sievert (Sv) (100-rem) dose established by 10 CFR 70.61(b)(1). For processes with effective engineered shielding, criticalities may actually produce doses below the intermediate consequences of 10 CFR 70.61. As stated in 10 CFR 70.61(d), such processes must nevertheless be subcritical for all normal and credible abnormal conditions, and primary reliance must be on prevention. This applies notwithstanding shielding or other mitigative features.

If needed, NUREG/CR-6410 provides methods for estimating the magnitudes of criticality events that can be applied for workers or members of the public at varying distances from the event.

iii. Likelihood information in the ISA Summary is acceptable to show compliance with 10 CFR 70.61, provided that the following conditions are met:

- The ISA Summary specifies the likelihood of each general type of accident sequence that could exceed the performance requirements of 10 CFR 70.61.

- The likelihoods are derived using an acceptable method described in the ISA Summary's section on methods.

- The likelihoods comply with acceptable definitions of the terms "unlikely" and "highly unlikely," as described in this SRP chapter. When interpreted as required accident frequencies, these terms refer to long-run average frequencies, not instantaneous values. That is, a system complies with the performance requirements of 10 CFR 70.61 as a long-run average. Otherwise, failure of any IROFS, even for a very short period, would violate the requirement, which is not the intent.

b. Management Measures. According to 10 CFR 70.65(b)(4), the ISA Summary must include a description of the management measures to be applied to IROFS, as well as information necessary to demonstrate compliance with the performance requirements of 10 CFR 70.61. Chapter 11 of this SRP provides detailed criteria for use in evaluating the adequacy of such management measures.

c. Criticality Monitoring. The regulation in 10 CFR 70.24 defines specific sensitivity and coverage requirements for criticality monitors. Chapter 5 of this SRP describes the acceptance criteria and review of information supporting a demonstration of compliance with 10 CFR 70.24.

Specific emergency preparations are also required by 10 CFR 70.24. Specifically, the application should provide information to demonstrate that the applicant's equipment and procedures are adequate to meet these requirements.

d. Requirements for New Facilities or New Processes at Existing Facilities. The baseline design criteria specified in 10 CFR 70.64 must be used, as applicable, for new facilities and new processes at existing facilities. If the application involves such new facilities or processes, the ISA Summary should explain how the design of the facility addresses each baseline design criterion. For deterministic design criteria such as double contingency, the process-specific information may be provided, along with the other process information in the ISA Summary. The application should also describe the design-basis events and safety parameter limits. In addition, the application should provide methods, data, and results of analysis showing compliance with these design bases for individual processes and facilities.

The regulation in 10 CFR 70.64 states that the design process must be founded on defense-in-depth principles and must incorporate, to the extent practicable, preference for engineered controls over administrative controls and reduction of challenges to IROFS. Because of this regulation, new facilities with system safety designs lacking defense-in-depth practices, consisting of purely administrative controls, or relying on IROFS that are frequently or continuously challenged, are not acceptable, unless the application provides a justification showing that alternatives to achieve the design criteria are not feasible.

(5) ISA Team Qualifications and ISA Methods. The ISA teams (10 CFR 70.62(c)(2)) and their qualifications as stated in the ISA Summary are acceptable if the following criteria are met:

 a. The ISA team has a leader who is formally trained and knowledgeable in the ISA methods chosen for the hazard and accident evaluations. In addition, the team leader should have an adequate understanding of all process operations and hazards under evaluation but should not be the responsible, cognizant engineer or expert for that process.

 b. At least one member of the ISA team has thorough, specific, and detailed experience in the type of process design under evaluation.

 c. The team has a variety of process design and safety experience in the particular safety disciplines relevant to hazards that could credibly be present in the process, including, if applicable, radiation safety, nuclear criticality safety, fire protection, and chemical safety disciplines.

 d. A manager provides overall administrative and technical direction for the ISA.

The description of the ISA methods is acceptable if the following criteria are met:

 a. Hazard Identification Method. The hazard identification method selected is considered acceptable if it meets the following criteria:

  i. The description includes a list of materials (radioactive, fissile, flammable, and toxic) and conditions that could result in hazardous situations (e.g., loss of containment of licensed nuclear material). The list should include maximum intended inventory amounts and the location of the hazardous materials at the facility.[1]

  ii. The method has determined potential interactions between materials or upset conditions that could result in hazardous situations where not normally present.

 b. Process Hazard Analysis Method. The process hazard analysis method is acceptable if it involves selecting one of the methods described in NUREG-1513

---

[1] At a minimum, the inventory list should include the following hazardous materials if present on site: ammonia, fines (uranium oxide dust, beryllium), flammable liquids and gases, fluorine, hydrofluoric acid, hydrogen, nitric acid, organic solvents, propane, uranium hexafluoride, and Zircaloy.

in accordance with the selection criteria established in that document. Methods not described in NUREG-1513 may be acceptable provided that they fulfill the following conditions:

i. Criteria are provided for their use for an individual process and are consistent with the principles of the selection criteria in NUREG-1513.

ii. The method adequately addresses all the hazards identified in the hazard identification task. If an identified hazard is eliminated from further consideration, such action is justified.

iii. The method provides reasonable assurance that the applicant can identify all significant accident sequences (including the IROFS used to prevent or mitigate the accidents) that could exceed the performance requirements of 10 CFR 70.61.[2]

iv. The method considers the interactions of identified hazards and proposed IROFS, including system interactions that could result in an accident sequence for which the consequences could exceed the performance requirements of 10 CFR 70.61.

v. The method addresses all modes of operation, including startup, normal operation, shutdown, and maintenance.

vi. The method addresses hazards resulting from process deviations (e.g., high temperature and high pressure), initiating events internal to the facility (e.g., fires or explosions), and hazardous credible external events (e.g., floods, high winds, earthquakes, and airplane crashes). The applicant provides justification for determinations that certain events are not credible and, therefore, not subject to the likelihood requirements of 10 CFR 70.61.

vii. The method adequately considers initiation of or contribution to accident sequences by human error through the use of human-systems interface analysis or other appropriate methods.

viii. The method adequately considers common mode failures and system interactions in evaluating systems that rely on redundant controls.

ix. The ISA Summary provides justification that the individual method would comply with conditions (ii) through (viii), above.

c. Consequence Analysis Method. The methods used for ISA consequence evaluation, as described in the ISA Summary, are acceptable, provided that the following conditions are met:

---

[2] The release of hazardous chemicals is of regulatory concern to the NRC only to the extent that such hazardous releases result from the processing of licensed nuclear material or have the potential to adversely affect radiological safety.

i.      The methods are consistent with the approaches described in NUREG/CR-6410.

ii.      The use of generic assumptions and data is reasonably conservative for the types of accidents analyzed.

d.      Likelihood Evaluation Method. The method for evaluating the likelihood of accident sequences, as described in the ISA Summary, is considered acceptable, provided that it meets the following conditions:

i.      The method clearly shows how each designated IROFS acts to prevent or mitigate the consequences (to an acceptable level) of the accident sequence being evaluated.

ii.      When multiple IROFS are designated for an accident sequence, the method considers the interaction of all such IROFS, as in a logic diagram or tabulation that accounts for the impact of redundancy, independence, and surveillance on the likelihood of occurrence of the accident.

iii.      The method has objective criteria for evaluating, at least qualitatively, the likelihood of failure of individual IROFS. When applicable, such likelihood criteria should include the means to limit potential failure modes, the magnitude of safety margins, the type of engineered equipment (active or passive) or human action that constitutes the IROFS, and the types and safety grading (if any) of the management measures applied to the IROFS.

iv.      The method evaluates the likelihood of each accident sequence as "unlikely," "highly unlikely," or neither, as defined by the applicant, in accordance with Section 3.4.3.2, item (9), of this chapter.

v.      For nuclear criticality accident sequences, the method evaluates compliance with 10 CFR 70.61(d). That is, even in a facility with engineered shielding to limit the consequences of nuclear criticalities, *preventive* controls must be in place that are sufficient to ensure that the process is subcritical for all credible abnormal conditions. Compared to unshielded processes, a moderately higher likelihood may be permitted in preventing such events, consistent with American National Standards Institute/American Nuclear Society (ANSI/ANS) Standard 8.10, "Criteria for Nuclear Criticality Safety Controls in Operations with Shielding and Confinement," reaffirmed in 2005. In particular, criticality cannot result from any credible failure of a single IROFS. In addition, potential criticality accidents must meet an approved margin of subcriticality for safety. Acceptance criteria for such margins are reviewed as programmatic commitments, but the ISA methods must consider, and the ISA Summary must document, the actual magnitude of those margins when they are part of the reason that the postulated accident sequence resulting in criticality is deemed highly unlikely.

Appendix A to this chapter provides an example of one acceptable method for evaluating likelihood that is based on a likelihood index. Appendix B offers additional guidance on acceptable methods for qualitative evaluation of likelihood. Appendix C discusses issues relating to the use of initiating event frequencies in demonstrating compliance with the likelihood requirements. Appendix D discusses acceptable ways for the ISA to address natural phenomena.

(6)      Descriptive List of All IROFS. The "list describing items relied on for safety" required by 10 CFR 70.62(c)(1)(vi) is acceptable, provided that it meets the following conditions:

     a.      The list includes all IROFS in the identified high- and intermediate-consequence accident sequences.

     b.      The description of the IROFS may include management measures applied to the IROFS (including the safety grading); should include the characteristics of its preventive, mitigative, or other safety function; and may include assumptions and conditions, such as safety limits or margins, if these are needed to understand how the item is capable of achieving compliance with 10 CFR Part 70, Subpart H.

The above acceptance criteria are explained in greater detail below.

     a.      The primary function of the list describing each IROFS is to document the safety basis of all processes in the facility. This list assists in ensuring that the items (IROFS) are not degraded without a justifying safety review. Thus, the key feature of this list is that it includes *all* IROFS. To be acceptable, no item, control, or control system of a process that is needed to show compliance with the safety performance requirements of the regulation may be omitted from this list (see 10 CFR 70.61(e)). However, sets of hardware or procedures that perform the same safety function may be referred to as a single set of IROFS and do not need to be individually identified. The list of IROFS may erroneously be incomplete in a number of ways: (1) an ineffective method of identifying accident sequences may have been used, (2) in applying the method to identify accidents something was overlooked, (3) a whole area or process subject to accidents was improperly screened out or simply omitted from the ISA, (4) IROFS were not applied to an identified accident, or (5) the list of accidents was incomplete because of incompleteness in the process design itself. The reviewer should attempt, in the horizontal slice review, to determine if any of these errors has occurred.

     b.      IROFS may be hardware with a dedicated safety function or hardware with a property that is relied on for safety. Thus, IROFS may be the dimension, shape, capacity, or composition of hardware. The ISA Summary need not provide a breakdown of hardware IROFS by component or identify all support systems. However, the ISA documentation maintained on site, such as system schematics and/or descriptive lists, should contain sufficient detail about items within a hardware IROFS, that it is clear to the reviewer and the applicant what structure, system, equipment, or component is included within the hardware IROFS' boundary and would, therefore, be subject to management measures specified by the applicant. Some examples of items within a hardware IROFS are detectors, sensors, electronics, cables, valves, piping, tanks, and dikes. In addition, ISA documentation should also identify essential utilities and support systems on which the IROFS depends to perform its intended function. Some examples of these are backup batteries, air supply, and steam supply. In some processes, the frequency of demands made on IROFS must be controlled or limited to comply with 10 CFR 70.61. In such processes, whatever features are needed to limit the frequency of demands are themselves IROFS.

c. The essential features of each IROFS should be described. Sufficient information should be provided about engineered hardware controls to permit an evaluation that, in principle, controls of this type will have adequate reliability. Because the likelihood of failure of items often depends on safety margins, descriptions of the safety parameter controlled by the item, the safety limit on the parameter, and the margin to true failure may be needed. For IROFS that are administrative controls, the nature of the action or prohibition involved must be described sufficiently to permit an understanding that, in principle, adherence to it should be reliable. Features of the IROFS that affect its independence from other IROFS, such as reliance on the same power supplies, should be indicated.

The description of each IROFS should identify its expected function, conditions needed for the IROFS to reliably perform its function, and the effects of its failure. The description of each IROFS within an ISA Summary should identify the management measures, such as maintenance, training, and configuration management that are applied to it. If a system of graded management measures is used, the grade applied to each control should be determinable from information in the ISA Summary. The reliability required for an IROFS is proportionate to the amount of risk reduction it is expected to supply. Thus, the quality of the management measures applied to an IROFS may be graded commensurate with the required reliability. The management measures should ensure that IROFS are designed, implemented, and maintained, as necessary, to be available and reliable to perform their function when needed. The degree of reliability and availability of IROFS ensured by these measures should be consistent with the evaluations of accident likelihoods. In particular, for redundant IROFS, all information necessary to establish the average vulnerable outage time is required in order to maintain acceptable availability. Otherwise, failures must be assumed to persist for the life of the facility. In particular, for IROFS whose availability is to be relied on, the time interval between surveillance observations or tests of the item should be stated, since restoration of a safe state cannot occur until the failure is discovered.

Table A-13 in Appendix A to this chapter is one example of a tabular description of IROFS meeting these criteria.

(7) Quantitative Standards for Chemical Consequences. The applicant's description in the ISA Summary of proposed quantitative standards used to assess consequences from acute chemical exposure to licensed material or chemicals incident to the processing of licensed material is acceptable, provided that the following criteria are met:

a. Unambiguous quantitative standards exist for each of the applicable hazardous chemicals that meet the criteria of 10 CFR 70.65(b)(7) on site, corresponding to, and consistent with, the quantitative standards in 10 CFR 70.61(b)(4)(i), 70.61(b)(4)(ii), 70.61(c)(4)(i), and 70.61(c)(4)(ii).

b.   The quantitative standard of 10 CFR 70.61(b)(4)(i) addresses exposures that could endanger the life of a worker. The applicant is appropriately conservative in applying the language "could endanger," so as to include exposures that could result in death for some workers, consistent with the methods used in the U.S. Environmental Protection Agency's acute exposure guidelines in Appendix A, "Table of Toxic Endpoints," to 40 CFR Part 68, "Chemical Accident Prevention Provisions."

c.   The quantitative standards for 10 CFR 70.61(b)(4)(ii) and 10 CFR 70.61(c)(4)(i) will correctly categorize all exposures that could lead to irreversible or other serious, long-lasting health effects to individuals. As with criterion (b) above, the standard selected should have appropriate conservatism.

d.   The quantitative standard for 10 CFR 70.61(c)(4)(ii) will correctly categorize all exposures that could cause mild transient health effects to an individual.

The NRC finds the use of the Emergency Response Planning Guidelines (ERPGs) established by the American Industrial Hygiene Association, the Acute Exposure Guideline Levels (AEGLs) established by the National Advisory Committee for Acute Guideline Levels for Hazardous Substances, and exposure limits established by the Occupational Safety and Health Administration or contained in International Organization for Standardization (ISO) standards to be acceptable. If the applicant does not use a published exposure standard, or if a chemical has an unknown exposure standard, the ISA Summary must describe how an alternative exposure standard was established for use in the ISA. The ISA Summary must list the actual exposure values for each chemical, specify the source of the data (e.g., ERPG, AEGL, ISO), and provide information or a reference supporting the claim that they meet the acceptance criteria stated above. (See also Section 6.4.3.1 of this SRP.)

(8)   List of Sole IROFS. The descriptive list in the ISA Summary that identifies all IROFS that are the sole item credited as such for demonstrating compliance with 10 CFR 70.61 is acceptable if it includes the following:

a.   descriptive title of the IROFS

b.   an unambiguous and clear reference to the process to which the item applies

c.   clear and traceable references to the description of the item as it appears in the full list of all IROFS and the list of accident sequences

(9)   Definitions of "Unlikely," "Highly Unlikely," and "Credible." The regulation in 10 CFR 70.65 requires that the applicant's ISA Summary must define the terms "unlikely," "highly unlikely," and "credible." The applicant's definitions of these terms are acceptable if, when used with the applicant's method of assessing likelihoods, they provide reasonable assurance that the performance requirements of 10 CFR 70.61 can be met. The applicant's *method* of likelihood evaluation and the *definitions* of the likelihood terms are closely related. Qualitative methods require qualitative definitions. Such a qualitative definition would identify the qualities of IROFS controlling an accident sequence that would qualify that sequence as "unlikely" or "highly unlikely."

An applicant may use quantitative methods and definitions for evaluating compliance with 10 CFR 70.61, but nothing in this SRP should be construed as an interpretation that such methods are required. The reviewer should focus on objective qualities and information provided concerning accident likelihoods.

As stated in 10 CFR 70.61, credible high-consequence events must be "highly unlikely." Thus, the meaning of the phrase "highly unlikely" is on a per-event basis. The same is true for the terms "unlikely" and "credible." Hence, applicant definitions should be on a per-event basis. The events referred to are occurrences of consequences, which are synonymous with the phrase "accident sequence" in this context. This is important to recognize, since an ISA may identify hundreds of potential accident sequences. Thus, the likelihood of each individual sequence must be quite low.

Acceptance Criteria for the Definition of "Credible"

The regulation in 10 CFR 70.65 requires that the applicant define the term "credible." This term is used in 10 CFR 70.61, which requires that all credible accident sequences for which the consequences could exceed the performance requirements of 10 CFR 70.61 must be controlled to be unlikely or highly unlikely, as appropriate. If an event is not credible, IROFS are not required to prevent or mitigate the event. Thus, to be "not credible" could be used as a criterion for exemption from use of IROFS. This raises a danger of circular reasoning. In the safety program embodied in Subpart H to 10 CFR Part 70, the "not credible" nature of an event must not depend on any facility feature that could credibly fail to function or be rendered ineffective as a result of a change to the system. Each facility feature that is needed to ensure that accident events are sufficiently unlikely is an IROFS. Management measures must offer high assurance, that such features are not removed or rendered ineffective during system changes. One cannot claim that a process does not need IROFS because it is "not credible" due to characteristics provided by some other controls or features of the plant that are not IROFS. Such an evaluation would be inconsistent with 10 CFR 70.61. However, although an accident sequence may not meet a definition of "not credible," it may meet the standards for "highly unlikely" or "unlikely" because of an infrequent external initiating event, without the use of IROFS. In such a case, IROFS are not necessary, but information is needed to show that the event does qualify as "highly unlikely" or "unlikely."

Any one of the following three independent acceptable sets of qualities could define an event as not credible:

(1)     An external event has a frequency of occurrence that can conservatively be estimated as less than once in a million years.

(2)     A process deviation consists of a sequence of many unlikely events or errors for which there is no reason or motive. In determining that there is no reason for such errors, a wide range of possible motives, short of intent to cause harm, must be considered. Complete ignorance of safe procedures is possible for untrained personnel, which should be considered a credible possibility. Obviously, no sequence of events should be categorized as not credible if it has actually occurred in any fuel cycle facility.

(3)     A convincing argument exists that, given physical laws, process deviations are not possible, or are extremely unlikely. The validity of the argument must not depend on any feature of the design or materials controlled by the facility's system of IROFS or management measures.

Such a demonstration of "not credible" must be convincing despite the absence of designated IROFS. Typically, this can be achieved only for external events known to be extremely unlikely.

Acceptance Criteria for Qualitative Definitions of Likelihood

If the applicant's definitions are qualitative, they are acceptable if they meet all of the following criteria:

*   They are reasonably clear and based on objective criteria.

*   They can reasonably be expected to consistently distinguish accidents that are "highly unlikely" from those that are merely "unlikely."

*   Their categorization of events as "highly unlikely" or "unlikely" yields results reasonably consistent with quantitative information and quantitative criteria such as those given in the example here.

The phrase "objective criteria" means the extent to which the method relies on specific identifiable characteristics of a process design, rather than subjective judgments of adequacy. Objective criteria are needed to achieve consistency. "Consistency" means the degree to which different analysts obtain the same results when they apply the method. This is important in maintaining an adequate standard of safety because the ISAs of future facility modifications may be performed by individuals not involved in conducting the initial ISA. An acceptable qualitative method of likelihood evaluation should yield results comparable to the examples of evaluation methods given in the appendices to this chapter.

Reliability and Availability Qualities

Qualitative methods of evaluating the likelihood of an accident sequence involve identifying the reliability and availability qualities of each of the events that constitute the sequence. The following lists of qualities are not necessarily complete, but they do contain many of the factors most commonly encountered. Some of these qualities relate to the characteristics of individual IROFS, such as the following examples:

*   safety margin in the controlled parameter, compared with process variation and uncertainty

*   whether the IROFS is an active engineered control, a passive engineered control, an administrative control, or an enhanced administrative control

*   the type and safety grading, if any, of management measures applied to the control

- fail-safe, self-announcing, or surveillance measures to limit downtime

- failure modes

- demand rate

- failure rate

Other reliability qualities relate to characteristics of the IROFS or system of IROFS that protect against the following accident sequences as a whole, among others:

- defense in depth
- degree of redundancy
- degree of independence
- diversity
- vulnerability to common-cause failure

Methods of likelihood evaluation and definitions of the likelihood terms "unlikely" and "highly unlikely" may mix qualitative and quantitative information. Certain types of objective quantitative information may be available concerning specific processes in a facility. Examples of such objective quantitative information include the following:

- reports of failure modes of equipment or violations of procedures recorded in maintenance records or corrective action programs

- the time intervals at which surveillance is conducted to detect failed conditions

- the time intervals at which functional tests or configuration audits are held

- for a fail-safe, monitored, or self-announcing IROFS, the time it takes to render the system safe

- demand rates (i.e., the frequency of the demands on an IROFS to perform) (some situations amount to effectively continuous demand)

Such items of quantitative information should be considered in evaluating the likelihood of accident sequences, even in purely qualitative evaluations. For example, knowing the value to which downtime is limited by surveillance can indicate that a system's availability is extremely high. For redundant systems, such high availability can virtually preclude concurrent independent failures of multiple IROFS.

Acceptance Criteria for Likelihood Indexing Methods

One acceptable definition for the likelihood terms "unlikely" and "highly unlikely" could be based on a risk-indexing method. The example in Appendix A to this chapter shows the use of such a method, which primarily relies on a qualitative evaluation of reliability and availability factors. In such methods, qualitative characteristics of the system of IROFS, such as those listed above, are used to estimate a quantitative likelihood index for each accident sequence. Then, the definitions of "highly unlikely" and "unlikely" would be acceptable limiting values of this likelihood index. For example, "highly unlikely" could be defined as "having a risk index value less than or equal to minus 5," and "unlikely" could be defined as "having a risk index value less than or equal to minus 4."

Acceptance Criteria for Purely Qualitative Methods

A purely qualitative method of defining "unlikely" and "highly unlikely" is acceptable if it incorporates all of the applicable reliability and availability qualities to an appropriate degree. For example, one statement of applicable qualities is double-contingency protection, the quality of a process design that incorporates sufficient factors of safety to require at least two unlikely, independent, and concurrent changes in process conditions before a criticality accident is possible.

Double contingency explicitly addresses several reliability and availability qualities:

- factors of safety: safety margins
- at least two: redundancy
- unlikely: low failure rate, low downtime of one of two controls
- concurrent: low downtime
- independent: independence
- process conditions: physical events, not virtual human errors

One acceptable definition of "highly unlikely" is a system of IROFS that possesses double-contingency protection, where each of the applicable qualities is present to an appropriate degree. For example, as implied by the modifier "at least" in the phrase "at least two unlikely, independent and concurrent changes," sometimes more than two-fold redundancy may be appropriate.

A qualitative method may also be proposed for defining "unlikely." Such a qualitative method might simply list various combinations of reliability qualities for a system of IROFS that would qualify as "unlikely." For example, a single high-reliability IROFS, such as an engineered hardware control with a high grade of applicable management measures, might qualify to be considered "unlikely to fail." Systems relying on administrative controls would normally have to use enhancing qualities, such as large safety margins and redundancy, to qualify as "unlikely to fail." A single simple administrative control, regularly challenged, and without any special safety margin or enhancement, where a single simple error would lead to an accident, would not qualify as "unlikely to fail." Likewise, two simple administrative controls without special margins or enhancements, particularly of their independence, would not normally qualify as "highly unlikely to fail."

Acceptance Criteria for Quantitative Definitions of Likelihood

An applicant may choose to provide quantitative definitions of the terms "unlikely" and "highly unlikely." One example of acceptable quantitative guidelines is given in the next paragraph. These guidelines serve two purposes. Specifically, these guidelines can be used as acceptance criteria for quantitative definitions of "highly unlikely" and "unlikely," if provided by an applicant.

## Quantitative Guidelines

A discussion of quantitative guidelines here does not imply that quantitative demonstration of compliance with 10 CFR 70.61 is required. The NRC has provided various guidance documents, including a strategic plan pertinent to ensuring that exposures of individuals to NRC-regulated hazards, such as radiation, are acceptably infrequent. For example, the NRC Strategic Plan has a performance goal of "no inadvertent nuclear criticalities." The quantitative guidelines given below for definitions of "highly unlikely" and "unlikely", as used in 10 CFR 70.61, were developed so as to be reasonably consistent with other relevant NRC guidance.

## Highly Unlikely

Among other considerations, the guideline for acceptance of the definition of "highly unlikely" has been derived as the highest acceptable frequency that is consistent with the performance goal of having no inadvertent nuclear criticality accidents. This guideline is thus applied in considering the 10 CFR 70.61 requirement that high-consequence events be highly unlikely, because such events may involve high radiation doses, as is often true for criticality accidents. To within an order of magnitude, this is taken to mean a definition that translates to a frequency limit of less than one such accident in the industry every 100 years. This results in a guideline limiting the frequency of any individual accident to $10^{-5}$ per event, per year. As the goal is to have no such accidents, the expectation is that most accidents would have frequencies substantially below this guideline when feasible.

## Unlikely

Intermediate-consequence events include significant radiation exposures to workers (those exceeding 0.25 Sv or 25 rem). The NRC has a strategic goal that there be no increase in the rate of such significant exposures. This guideline has been interpreted here to correspond to a range between $10^{-4}$ and $10^{-5}$ per event, per year.

## Quantitative Guidelines for Use with Acceptance Criteria

The applicant's quantitative definitions of the terms "unlikely" and "highly unlikely," as applied to individual accident sequences identified in the ISA, are acceptable to show compliance with 10 CFR 70.61 if they are reasonably consistent with the following quantitative guidelines:

| Likelihood Term of 10 CFR 70.61 | Guideline |
|---|---|
| Unlikely | Less than $10^{-4}$ per event, per year |
| Highly Unlikely | Less than $10^{-5}$ per event, per year |

The stated quantitative guidelines are used to define the *largest* likelihood values that would be acceptable limits. Definitions based on lower limits are also acceptable. Note that the word "unlikely" as it appears in 10 CFR 70.61(c) does not have the same meaning as it does in the definition of double contingency. (See Chapter 5 of this SRP.)

## 3.5 Review Procedures

Organization of the reviews addressed by this chapter of the SRP will differ depending on the scope of the documents submitted. For a license application, renewal, or amendment application containing a new or revised chapter addressing the applicant's safety program and ISA commitments, there may be only a primary ISA reviewer. However, for an initial ISA Summary submittal, specialists in the various safety disciplines and management measures will assist the primary ISA reviewer. An ISA Summary update submitted as part of an amendment for a process that has hazards in multiple disciplines would also require a team approach. In general, a primary ISA reviewer will evaluate generic methods, risk, and reliability criteria used in the ISA and generic information about individual processes. Assisting this primary reviewer will be secondary reviewers, who will evaluate selected individual accidents and advise on the completeness of the accident list for specific safety disciplines.

### 3.5.1 Acceptance Review

For review of safety program commitments, including commitments pertaining to the ISA and ISA Summary, in a renewal or amendment application, the primary ISA reviewer will conduct a review to determine if the submittal contains appropriate information addressing each of the areas of review identified in Section 3.3.1 of this chapter. If the application does not contain sufficient information to permit a safety evaluation, the application will not be accepted for review.

For an ISA Summary, the primary ISA reviewer will also conduct an acceptance review to determine whether the document submitted contains sufficient information addressing the areas of review noted in Section 3.3.2, including specifically each of the elements required by 10 CFR 70.65(b), to permit an evaluation of safety for compliance with the regulations. If sufficient information is not present, the ISA Summary will not be accepted for review.

### 3.5.2 Safety Evaluation

#### 3.5.2.1 Evaluation of Safety Program and Integrated Safety Analysis Commitments

The reviewer examines the descriptions and commitments to program elements in the application or other documents for the areas of review described in Section 3.3.1 to ascertain whether the program elements are sufficient to meet the acceptance criteria of Section 3.4.3.1. The ISA reviewer must coordinate his or her review with reviews being conducted under other chapters of this SRP.

#### 3.5.2.2 Evaluation of ISA Summary

A team consisting of a primary reviewer together with specialists in each category of accidents would normally perform an evaluation of the ISA Summary to determine if it meets the acceptance criteria of Section 3.4.3.2. These categories of accidents depend on the facility, but in general, they are nuclear criticality, fires, chemical accidents, and radiological accidents. If external event analysis is complex, specialists may be employed to review these separately, as well. The primary ISA reviewer would normally evaluate the acceptability of the generic elements of the ISA Summary, such as site and facility descriptions, ISA methods, criteria, and consequence and likelihood definitions. However, each specialist should also review these elements to obtain information in support of his or her own evaluations.

In contrast to these generic ISA elements, process-specific information is needed by, and must be acceptable to, all of the specialists. Thus, all team members should evaluate the process descriptions in the ISA Summary.

Separate specialists for each category of accidents (i.e., nuclear criticalities, fires, radiological releases, and chemical accidents) should undertake the reviews of accident sequence descriptions and the likelihood and consequence information showing compliance with 10 CFR 70.61. As indicated in Appendix A to this chapter, one acceptable format for the ISA Summary is to separately tabulate or give logic diagrams for accident sequences in each accident category.

After a preliminary team review of the ISA Summary, the team should visit the facility to become familiar with the three-dimensional geometry of process equipment, to review components of the ISA, and to address any issues that arose during review of the ISA Summary.

To select a subset of the accident sequences reported in the ISA Summary for more detailed review, the reviewer should look at the applicant's tabulation of high- and intermediate-consequence accident sequences and the types of IROFS designated for each. High-consequence accident sequences protected by administrative controls should be examined very carefully, whereas intermediate consequence accident sequences protected by redundant passive engineered controls warrant less scrutiny.

To select specific accident sequences and IROFS for more detailed evaluation, the reviewer should evaluate potential accidents using information supplied in the ISA Summary. The applicant's method for identifying and establishing the consequences and likelihood of an accident sequence may provide information sufficient for this purpose. The NRC reviewer may evaluate the accidents using qualitative screening criteria analogous to those in Table A-6 in Appendix A to this chapter. Other, more rigorous reliability or consequence analyses may be performed as deemed necessary. On the basis of this analysis, accidents will be categorized. The reviewer may elect to examine in greater detail the engineered and administrative controls for accidents in the category of highest consequences. While on site, the reviewer should also select for specific evaluation a small sample of accident sequences determined by the applicant either to result in less than intermediate consequences or to be not credible.

From the list of the IROFS, the reviewer should categorize IROFS so that similar items are grouped together. The reviewer should then ensure that he or she has fully understood one or more prototype IROFS selected from each category. For these selected prototypes, the reviewer may, if necessary, request additional information needed to completely understand a particular IROFS. For complex processes, the reviewer may need to visit the facility to reach an adequate understanding of how the IROFS work for the process.

### 3.5.2.3 Onsite Integrated Safety Analysis Review

The reviewer should plan on visiting the applicant's facility at least once as part of the application review process. This visit should be scheduled after the applicant's ISA Summary has received a preliminary review. The visits will enable the reviewer to confirm through detailed examination of the ISA and ISA documentation that the ISA methods were selected and applied in a reasonable and thorough manner to all facility processes, that all credible high- and intermediate-consequence accident sequences were correctly identified, that accident sequence consequences and likelihoods were reasonably determined, and that appropriate IROFS and supporting management measures have been proposed. By means of a "horizontal" review and several "vertical" slice reviews (defined below) of processes selected by the reviewer, the NRC staff can establish the completeness and adequacy of the applicant's ISA method. The reviewer may use the ISA documentation to perform independent evaluations of process hazards and accident sequences using methods selected from NUREG-1513, Appendix A to this SRP chapter, or other NRC guidance.

The reviewer should not attempt a comprehensive, all-encompassing review of every facility process and every accident sequence on the site visit. Rather, the reviewer should use the site visit to confirm the appropriateness and adequacy of the applicant's ISA method and the completeness of the ISA and accuracy of analysis of accident sequences by means of a horizontal review and several vertical slice reviews of selected processes. The site visit will also afford the reviewer an opportunity to seek answers to questions from the applicant (or possibly the ISA team) that may have arisen in the preliminary review of the ISA Summary.

The following discusses each of the three facets of the onsite ISA review:

(1)     ISA Methods Review

The purpose of the ISA methods review is two-fold: (a) to ensure that the applicant selected appropriate ISA methods for each facility process and (b) to ensure that the methods were correctly applied in conducting the ISA. The ISA Summary should describe the ISA methods and give a few examples of the application of the ISA methods. The ISA methods review should answer any questions that a reviewer may have concerning ISA methods and procedures after completion of the preliminary review of the ISA Summary. In reviewing process-specific information in the ISA Summary and ISA documentation maintained on site, the reviewer should select a few processes and accident sequences to examine the adequacy of the selected ISA methods and their application. The reviewer should examine any procedures, checklists, or guidance documents that the applicant may have on site as guidance to ISA team members to ensure a complete understanding of the applicant's ISA methods. The reviewer should then examine the ISA documentation, including the selected processes and accident sequences, showing how the ISA methods were applied as part of the horizontal and vertical slice reviews discussed below.

(2)     Horizontal Review

The basic purpose of the horizontal review is to ensure completeness of the ISA of facility processes. This does not require an absolute checkoff of ISA documentation against the full list of processes to be covered, but it does mean that a substantial fraction of the processes should receive a brief examination.

The reviewer should consult the ISA and ISA documentation to answer questions or to resolve outstanding issues resulting from the preliminary review of the ISA Summary. In particular, the reviewer should examine safety information that is not included in the ISA Summary. For example, ISA documentation related to hardware IROFS, such as system schematics and/or descriptive lists, should contain sufficient detail about hardware IROFS so that it is clear to the reviewer what components (such as cables, detectors, alarms, valves, and piping) are included within the boundary of the hardware IROFS system and would therefore be subject to management measures specified by the applicant. In addition, such documentation should also identify support systems (such as backup batteries, air supply, and steam supply) on which the IROFS depends to perform its intended function. The reviewer should also examine a few processes to confirm that all accident sequences were considered and that the ISA Summary includes those having potential consequences exceeding the performance requirements of 10 CFR 70.61.

(3)    Vertical Slice Review

The purpose of the vertical slice review is to examine the application of the ISA methods to a selected subset of facility processes. For this subset of facility processes, the reviewer should examine the underpinnings of calculations, conclusions, and the design of safety programs that result from the ISA, as well as safety information that is not identified in the ISA Summary. The reviewer should examine accident sequences for this subset of processes to determine the adequacy of the applicant's consequence and likelihood determinations. In addition, the reviewer should examine the appropriateness and robustness of designated IROFS and the suitability of proposed management measures.

The ISA Summary will have categorized accidents according to their consequences, likelihoods, and IROFS. The reviewer should select a subset of processes for vertical slice review of these categories. The subset should include accident sequences with relatively high levels of consequence and likelihood and accident sequences for which IROFS of different types and relatively low robustness are designated. For ISAs where the index method of Appendix A is used, and where the index scoring for all accident sequences is readily available to the reviewer, in principle, these index scores could be used to establish sequences of relatively higher risk. However, if the ISA declares as IROFS only a set of controls that are minimally necessary to demonstrate compliance with 10 CFR 70.61 likelihood requirements, then such index scores would be misleading. Instead, in selecting processes or sequences for the vertical slice reviews, one may need to use other objective qualities of the processes. For example, the selection might be based on experience or potential consequences as in (1) criticality accidents in solution systems, solvent extraction process upsets, or using plutonium or high-enriched uranium or (2) chemical processes involving large quantities of toxic chemicals that are highly reactive, flammable, or volatile, or are exceptionally toxic. Vertical slice reviews should examine processes for which less robust IROFS are designated (e.g., those with greater reliance on administrative rather than engineered controls). Again, if only a minimal set of IROFS is declared, it may be supported by more robust controls that are not IROFS and hence are not documented in the ISA Summary. Still, a review of sets of IROFS that are purely administrative, or are otherwise known from experience to be unreliable, may be advisable.

While on site, the reviewer may confirm the adequacy of sample accident analyses that the applicant included in the ISA Summary. However, the reviewer should focus on processes and/or accident sequences that were not included as sample accident analyses in the ISA Summary to ensure the completeness of the ISA.

The vertical slice review should address any specific questions the reviewer may have related to the ISA methods. If the applicant's methods are evaluated as effective in these selected cases, there is greater assurance that they will be effective for other processes. If questions or weaknesses are discovered that may be generic, the reviewer may have to perform vertical slice analyses on several additional processes. However, a specific question on the ISA of one process may not imply that there is a generic question requiring further examination. The purpose of the vertical slice reviews is not complete verification of ISA implementation.

The total number of vertical slice reviews to be conducted will depend on the facility's total number of accident sequences for which the consequences could exceed the performance requirements of 10 CFR 70.61, the diversity of the types of processes and types of IROFS at the facility, and the results of initial reviews of the ISA Summary and the horizontal and vertical slice reviews. For most fuel fabrication facilities, the reviewer should plan on conducting vertical slice reviews for 5 to 10 processes significant to nuclear criticality safety, 1 to 3 fire-significant processes, and 1 to 3 chemical/radiological/environmental-significant processes. However, if the initial reviews of the ISA Summary and the horizontal and vertical slice reviews identify significant issues, then additional vertical slice reviews may be warranted. Ultimately, to approve the ISA and ISA program, the reviewer must attain reasonable assurance that the applicant has implemented them in compliance with the regulations.

Each vertical slice review should include (1) familiarization of the reviewer with the safety design of the selected process and (2) examination of all onsite documentation related to the ISA of that process. If the content of the documentation leaves certain issues unclear, interviews with facility personnel may be necessary. The review should focus on the onsite information that is not provided in the ISA Summary but is key to determining compliance with 10 CFR 70.61 requirements.

Following the horizontal and vertical slice reviews, if outstanding questions remain about compliance with the performance requirements of 10 CFR 70.61, the reviewer may conduct an independent evaluation using appropriate methods selected from NUREG-1513, Appendix A to this chapter, or other agency guidance. The purpose of such an independent review is to identify strengths and weaknesses in the applicant's ISA methods or implementation practices, not simply to check compliance in this one case.

The reviewer should take care to document findings and evaluations made during this process.

## 3.6 Evaluation Findings

In general, the review findings should state that the requirements of 10 CFR 70.64 for a new facility, 10 CFR 70.65 for content, and 10 CFR 70.66, "Additional Requirements for Approval of License Application," have or have not been met, and the reasons for this finding. A finding statement should follow the evaluation of each specific area of review, stating how and why the information submitted in that area complies with the related regulatory requirement, if it does so. Specifically, the findings in the safety evaluation report should state conclusions of the following types:

- general conclusion resulting from the reviewer's evaluation of safety program commitments:

    The NRC staff concludes that the applicant's safety program, if established and maintained pursuant to 10 CFR 70.62, is adequate to provide reasonable assurance that IROFS will be available and reliable to perform their intended safety function when needed and in the context of the performance requirements of 10 CFR 70.61.

General findings for each of the areas of review should state how the applicant's information demonstrates compliance with the acceptance criteria of Section 3.4.3.1. If the reviewer finds that the acceptance criteria are not met and the applicant is not in compliance with the regulations, then the situation must be rectified before approval can be given. If the applicant has submitted an adequate explanation of an alternative way of complying with the regulations, the NRC's safety evaluation report should contain a finding that the alternative is acceptable to meet the basic regulatory requirement addressed.

- general conclusions resulting from the staff's evaluation of an ISA Summary:

> Many hazards and potential accidents can result in unintended exposure of persons to radiation, radioactive materials, or toxic chemicals incident to the processing of licensed materials. The NRC staff finds that the applicant has performed an ISA to identify and evaluate those hazards and potential accidents as required by the regulations. The NRC staff has reviewed the ISA Summary and other information and finds that it provides reasonable assurance that the applicant has identified IROFS and established engineered and administrative controls to ensure compliance with the performance requirements of 10 CFR 70.61. Specifically, the NRC staff finds that the ISA results, as documented in the ISA Summary, provide reasonable assurance that the IROFS, the management measures, and the applicant's programmatic commitments will, if properly implemented, make all credible intermediate-consequence accidents unlikely, and all credible high-consequence accidents highly unlikely.

Findings should be made concerning any specific requirement in 10 CFR Part 70 that addresses the nine elements in the ISA Summary. In particular, these findings should include statements concerning compliance with the requirements of 10 CFR 70.64 (regarding new facilities and new processes at existing facilities).

The review should result in findings concerning the compliance of specific processes with the requirements of 10 CFR 70.61, or other parts of the regulation, for those processes that receive specific detailed review. However, such findings should be limited to a finding of reasonable assurance that a process having the IROFS described in the ISA Summary is capable of meeting the requirements if properly implemented, operated, and maintained.

## 3.7 <u>References</u>

American Institute of Chemical Engineers, "Guidelines for Hazard Evaluation Procedures, Second Edition with Worked Examples," New York, September 1992.

American National Standards Institute/American Nuclear Society, "Nuclear Criticality Safety in Operations with Fissionable Materials Outside Reactors," ANSI/ANS-8.1-1983, 1983.

*U.S. Code of Federal Regulations*, Title 10, "Energy," Part 70, "Domestic Licensing of Special Nuclear Material."

U.S. Department of Commerce, Bureau of the Census, "Statistical Abstract of the United States," Table No. 688, 1995.

U.S. Nuclear Regulatory Commission, "Integrated Safety Analysis Guidance Document," NUREG-1513, 1995.

*U.S. Code of Federal Regulations*, Title 40, "Protection of Environment," Part 68, "Chemical Accident Prevention Provisions," Appendix A, "Table of Toxic Endpoints."

U.S. Nuclear Regulatory Commission, "Nuclear Fuel Cycle Facility Accident Analysis Handbook," NUREG/CR-6410, March 1998.

U.S. Nuclear Regulatory Commission, "Integrated Safety Analysis Guidance Document," NUREG-1513, May 2001.

U.S. Nuclear Regulatory Commission, "Chemical Process Safety at Fuel Cycle Facilities," NUREG-1601, August 1997.

*U.S. Code of Federal Regulations*, Title 29, "Labor," Section 1910.100, Chapter XVII, "Occupational Safety and Health Administration:

# APPENDIX A

# EXAMPLE PROCEDURE FOR ACCIDENT SEQUENCE EVALUATION

This appendix provides the U.S. Nuclear Regulatory Commission (NRC) reviewer with an example of one method of evaluating accident sequences for compliance with the likelihood requirements of Title 10 of the *Code of Federal Regulations* (10 CFR) 70.61, "Performance Requirements." It employs a semiquantitative risk index method for categorizing accident sequences in terms of their likelihood of occurrence and their consequences of concern. The risk index method framework will enable the applicant to identify, and the NRC reviewer to confirm, which accident sequences have consequences that exceed the performance requirements of 10 CFR 70.61 and, therefore, require designation of items relied on for safety (IROFS) and supporting management measures. The integrated safety analysis (ISA) summary should include descriptions of these general types of higher consequence accident sequences.

This appendix presents an example of how the risk index method can be applied to a uranium powder blender. It describes one method of evaluating compliance with the consequence and likelihood performance requirements of 10 CFR 70.61. The method is intended to permit any available quantitative information to be considered. For consistency, the NRC reviewer's approach could also include assigning quantitative values to any qualitative likelihood assessments made by an applicant, since likelihoods are inherently quantitative. This method should not be interpreted as a requirement that an applicant use quantitative evaluation. However, evaluation of a particular accident should be consistent with any facts available, which may include quantitative information concerning the availability and reliability of IROFS involved.

This appendix is not a "format and content guide" for either the ISA or the ISA Summary. It simply presents one method of analysis and categorization of credible accident sequences for facility processes. The method described in this appendix uses both qualitative and quantitative criteria for evaluating frequency indices of safety controls. These criteria for assigning indices, particularly the descriptive criteria provided in some tables of this appendix, are intended to be examples, not universal criteria. It is preferable that each applicant develop such criteria based on particular types of IROFS and management measure programs. The applicant should modify and improve such criteria as insights are gained during performance of the ISA.

If the applicant evaluates accidents using a different method, the method should produce similar results in terms of how accidents are categorized. The method should be regarded as a screening method, not as a definitive method of proving the adequacy or inadequacy of the IROFS for any particular accident. Because methods can rarely be universally valid, an evaluation using other methods may be justified for individual accidents for which this method does not appear applicable. The method does have the benefit that it evaluates, in a consistent manner, the characteristics of IROFS used to limit accident sequences. This will permit identification of accident sequences with defects in the combination of IROFS used. Such IROFS can then be further evaluated or improved to establish adequacy. The procedure also ensures the consistent evaluation of similar IROFS by different ISA teams. Sequences or IROFS that are risk significant and are evaluated as marginally acceptable are good candidates for more detailed evaluation by the applicant and the reviewer.

For each sequence, the tabular accident summary resulting from the ISA should identify the engineered or administrative IROFS that must fail to allow the occurrence of consequences that exceed the levels identified in 10 CFR 70.61. Chapter 3 of this Standard Review Plan (SRP) specifies acceptance criteria for these IROFS and for meeting the performance requirements of 10 CFR 70.61. These criteria require that IROFS be sufficiently unlikely to fail. However, the acceptance criteria do not explicitly mandate any particular method for assessing likelihood. The purpose of this appendix is to provide an example of an acceptable method to perform this evaluation of likelihood.

## A.1 Risk Matrix Development

### Consequences

The regulation in 10 CFR 70.61 specifies two categories for accident sequence consequences: "high consequences" and "intermediate consequences." Implicitly, there is a third category for accidents that produce consequences less than "intermediate." This category will be referred to as "low-consequence" accident sequences. The primary purpose of process hazard analysis (PHA) is to identify all uncontrolled and unmitigated accident sequences. These accident sequences can then be categorized into one of these three consequence categories (high, intermediate, or low) based on their predicted radiological, chemical, and/or environmental impacts. Although the subsequent ISA analysis focuses only on those accident sequences having high or intermediate consequences, by examining low-consequence events identified and tabulated in the ISA, the reviewer can evaluate the completeness of the PHA and ISA analyses. Table A-1 presents the radiological and chemical consequence severity limits of 10 CFR 70.61 for each of the three accident consequence categories.

### Table A-1 Consequence Severity Categories Based on 10 CFR 70.61

|                                         | Workers                                                                                 | Offsite Public                                                                                     | Environment                                                                          |
|-----------------------------------------|-----------------------------------------------------------------------------------------|---------------------------------------------------------------------------------------------------|--------------------------------------------------------------------------------------|
| **Category 3 High Consequence**         | *RD > 1 sievert (Sv) (100 rem) **CD = endanger life                                      | RD > 0.25 Sv (25 rem) 30 milligrams (mg) sol U intake CD = long-lasting health effects            |                                                                                      |
| **Category 2 Intermediate Consequence** | 0.25 Sv (25 rem) < RD ≤1 Sv (100 rem) CD = long-lasting health effects                   | 0.05 Sv (5 rem) < RD ≤0.25 Sv (25 rem) CD = mild transient health effects                         | Radioactive release > 5,000 x Table 2 of 10 CFR Part 20, Appendix B                  |
| **Category 1 Low Consequence**          | Accidents with lower radiological and chemical exposures than those above in this column | Accidents with lower radiological and chemical exposures than those above in this column           | Radioactive releases producing lower effects than those referenced above in this column |

\* RD = Radiological Dose
\*\* CD = Chemical Dose

### Likelihood

The regulation in 10 CFR 70.61 also specifies the permissible likelihood of occurrence of accident sequences of different consequences. High-consequence accident sequences must be "highly unlikely" and intermediate-consequence accident sequences must be "unlikely." Implicitly, accidents in the low-consequence category can have a likelihood of occurrence less than "unlikely" or simply "not unlikely." Table A-2 shows the likelihood of occurrence limits of 10 CFR 70.61 for each of the three likelihood categories.

### Table A-2 Likelihood Categories Based on 10 CFR 70.61

|  | Qualitative Description |
|---|---|
| **Likelihood Category 1** | Consequence Category 3 accidents must be "highly unlikely." |
| **Likelihood Category 2** | Consequence Category 2 accidents must be "unlikely." |
| **Likelihood Category 3** | Consequence Category 1 accidents may be "not unlikely." |

## Risk Matrix

The three categories of consequence and likelihood can be displayed as a 3x3 risk index matrix. By assigning a number to each category of consequence and likelihood, a qualitative risk index can be calculated for each combination of consequence and likelihood. The risk index equals the product of the integers assigned to the respective consequence and likelihood categories. Table A-3 illustrates the risk index matrix, along with computed risk index values. The shaded blocks identify accidents for which the consequences and likelihoods yield an unacceptable risk index and to which IROFS must be applied.

### Table A-3 Risk Matrix with Risk Index Values

| Severity of Consequences | Likelihood of Occurrence | | |
|---|---|---|---|
|  | Likelihood Category 1 Highly Unlikely (1) | Likelihood Category 2 Unlikely (2) | Likelihood Category 3 Not Unlikely (3) |
| **Consequence Category 3 High (3)** | Acceptable Risk 3 | Unacceptable Risk 6 | Unacceptable Risk 9 |
| **Consequence Category 2 Intermediate (2)** | Acceptable Risk 2 | Acceptable Risk 4 | Unacceptable Risk 6 |
| **Consequence Category 1 Low (1)** | Acceptable Risk 1 | Acceptable Risk 2 | Acceptable Risk 3 |

The risk indices can initially be used to examine whether the consequences of an uncontrolled and unmitigated accident sequence (i.e., without any IROFS) could exceed the performance requirements of 10 CFR 70.61. If the performance requirements could be exceeded, the applicant must designate IROFS to prevent the accident or to mitigate its consequences to an acceptable level. A risk index value less than or equal to 4 means that the accident sequence is acceptably protected against and/or mitigated. If the applicant provides this risk index in the ISA and ISA Summary, the reviewer can quickly scan these data to confirm that each accident sequence meets the performance requirements of 10 CFR 70.61.

If the risk index of an uncontrolled and unmitigated accident sequence exceeds 4, the likelihood of the accident must be reduced through designation of IROFS. In this risk index method, the likelihood index for the uncontrolled and unmitigated accident sequence is adjusted by subtracting a score corresponding to the type and number of IROFS that have been designated. Table A-4 lists the qualitative scores assigned to the four types of IROFS.

Reviewers should note that the qualitative scores assigned in Table A-4 are for illustrative purposes only. IROFS meeting the criteria for a particular score in Table A-4 could have a wide range of availability or reliability. Such coarse criteria are useful for screening purposes, but when the total evaluated likelihood score for an accident sequence lies near the acceptance guideline value, a more careful evaluation should be done. Such evaluations should consider the management measures applied to all the reliability and availability qualities of the IROFS, or system of IROFS, protecting against the accident, as explained in the likelihood acceptance criteria of Section 3.4.3.2 of this SRP.

### Table A-4  Qualitative Categorization of IROFS

| Numeric Value | Description of IROFS |
|---|---|
| 1 | Protection by a single trained operator with adequate response time **(Administrative IROFS)** |
| 2 | Protection by a single, active engineered IROFS, functionally tested on a regular basis **(Active Engineered IROFS)** |
| 3 | Protection by a single, passive engineered IROFS, functionally tested on a regular basis, or by an active engineered IROFS with a trained operator for backup **(Passive Engineered IROFS or Combined Engineered and Administrative IROFS)** |
| 4 | Protection by two independent and redundant engineered IROFS, as appropriate, functionally tested on a regular basis **(Combination of Two Active or Passive Engineered IROFS)** |

To demonstrate compliance with the performance requirements of 10 CFR 70.61, the ISA should assign a consequence category to each identified accident sequence. The likelihood of occurrence of those accident sequences identified as high- or intermediate-consequence events must then be assigned to one of the three likelihood categories. To be acceptable, the controlled and/or mitigated accident consequences and likelihoods must have valid bases, and the applicant must include the bases for all general types of high- and intermediate-consequence accident sequences in the ISA Summary.

## A.2 Consequence Category Assignment

Categorization of an accident sequence as a high-consequence event or an intermediate-consequence event, or neither, is based on the estimated consequences of prototype accidents. Although accident consequences can be determined by actual calculations, calculations need not be performed for each individual accident sequence listed for a process. Accident consequences may also be estimated by comparison to similar events for which reasonably bounding conservative calculations have been made. Categorization also requires consideration of acute chemical exposures that an individual could receive from licensed material or hazardous chemicals incident to the processing of licensed material. The applicant must select appropriate acute chemical exposure data and relate these data to the performance requirements of 10 CFR 70.61(b)(4) and (c)(4). This appendix uses the Acute Exposure Guideline Level (AEGL) and Emergency Response Planning Guideline (ERPG). AEGL-3 and ERPG-3 exposure levels are life threatening.

**Consequence Category 3 (High Consequences)** includes accidents resulting in any consequence specified in 10 CFR 70.61(b). These include (1) acute worker exposures of (a) radiation doses greater than 1 Sv (100 rem) total effective dose equivalent (TEDE), and (b) chemical exposures that could endanger life (above AEGL-3 or ERPG-3), and (2) acute exposures to members of the public outside the controlled area to (a) radiation doses greater than 0.25 Sv (25 rem) TEDE, (b) soluble uranium intakes greater than 30 mg, and (c) chemical exposures that could lead to irreversible or other serious long-lasting health effects (exceeding AEGL-2 or ERPG-2). An unshielded nuclear criticality would normally be considered a high-consequence event because of the potential for producing a high radiation dose to a worker.

**Consequence Category 2 (Intermediate Consequences)** includes accidents resulting in any consequence specified in 10 CFR 70.61(c). These include (1) acute exposures of workers to (a) radiation doses between 0.25 Sv (25 rem) and 1 Sv (100 rem) TEDE, and (b) chemical exposures that could lead to irreversible or other serious long-lasting health effects above AEGL-2 or ERPG-2), and (2) acute exposures of members of the public outside the controlled area to (a) radiation doses between 0.05 Sv (5 rem) and 0.25 Sv (25 rem) TEDE, (b) chemical exposures that could cause mild transient health effects (exceeding AEGL-1 or ERPG-1), and (3) release of radioactive material outside the restricted area that would, if averaged over a 24-hour period, exceed 5,000 times the values specified in Table 2 of Appendix B, "Annual Limits on Intake (ALIs) and Derived Air Concentrations (DACs) of Radionuclides for Occupational Exposure; Effluent Concentrations; Concentrations for Release to Sewerage," to 10 CFR Part 20, "Standards for Protection against Radiation."

**Consequence Category 1 (Low Consequences)** includes accidents with potential adverse radiological or chemical consequences but at exposures less than Categories 3 and 2.

Table A-5 shows this system of consequence categories.

**Table A-5 Consequence Severity Categories Based on 10 CFR 70.61**

|  | Workers | Offsite Public | Environment |
|---|---|---|---|
| **Category 3 High Consequence** | *RD > 1 Sv (100 rem) **CD > AEGL-3, ERPG-3 | RD > 0.25 Sv (25 rem) 30 mg sol U intake CD > AEGL-2, ERPG-2 | |
| **Category 2 Intermediate Consequence** | 0.25 Sv (25 rem) < RD ≤1 Sv (100 rem) AEGL-2, ERGP-2 < CD ≤AEGL-3, ERPG-3 | 0.05 Sv (5 rem) < RD ≤0.25 Sv (25 rem) AEGL-1, ERGP-1 < CD ≤AEGL-2, ERPG-2 | Radioactive release > 5,000 x Table 2 in Appendix B to 10 CFR Part 20 |
| **Category 1 Low Consequence** | Accidents with lower radiological and chemical exposures than those above in this column | Accidents with lower radiological and chemical exposures than those above in this column | Radioactive releases with lower effects than those referenced above in this column |

\*  RD = Radiological Dose
\*\*CD = Chemical Dose

The applicant should document the bases for bounding calculations of the consequence assignment in the ISA Summary submittal. NUREG/CR-6410, "Nuclear Fuel Cycle Facility Accident Analysis Handbook," issued March 1998, describes valid methods and data that the applicant or staff may use for confirmatory evaluations.

## A.3 Likelihood Category Assignment

An assignment of an accident sequence to a likelihood category is acceptable if it is based on the record of occurrences at the facility, the record of failures of IROFS at the facility, applicable event data for similar systems, objective qualitative criteria governing system failure rates and availability, or other methods that have objective validity. Because sequences leading to accidents often involve multiple failures, the likelihood of the whole sequence will depend on the frequencies of initiating events and failure likelihoods of engineered and administrative IROFS. The method of likelihood assignment used in this appendix relies on the expert engineering judgment of the analyst and includes assessment of the number, type, independence, and observed failure history of designated IROFS. Engineered and administrative IROFS, even those of the same types, have a wide range of reliability. By requiring explicit consideration of most of the underlying events and factors that significantly affect the likelihood of the accident and explicit criteria for assigning likelihood, greater consistency in assigning likelihood to accident sequences across different systems within a facility and among different applicants should be possible.

This section provides one example of a set of acceptable semiquantitative risk guidelines for determining compliance with the likelihood requirements of 10 CFR 70.61 when using methods of evaluation that are either quantitative or use the risk index method outlined in this appendix.

The performance criteria of 10 CFR 70.61 are formulated in terms of likelihood limits on each separate event sequence. The example guidelines given in Table A-6 are based on the acceptance criteria guidance on likelihood definitions given in Section 3.4.3.2 of this SRP.

**Table A-6  Example Likelihood Index Limit Guidelines**

|  | Likelihood Category | Event Frequency Limits* | Risk Index Limits |
|---|---|---|---|
| **Not Unlikely** | 3 | more than $10^{-4}$ per event, per year | > -4 |
| **Unlikely** | 2 | between $10^{-4}$ and $10^{-5}$ per event, per year | -4 to -5 |
| **Highly Unlikely** | 1 | less than $10^{-5}$ per event, per year | $\leq$-5 |

Any risk or risk index method of likelihood evaluation using criteria as simple as those provided in the example method in this appendix should not be relied on exclusively in deciding the acceptability of the likelihood of a given event sequence. Consideration of qualitative criteria, such as degree of defense in depth or independence of controls, may be used to alter decisions based on the example of simple semiquantitative criteria presented here.

## A.4  Assessing Effectiveness of Items Relied on for Safety

The risk of an accident sequence is reduced through application of different numbers and types of IROFS. By either reducing the likelihood of occurrence or by mitigating the consequences, IROFS can reduce the overall resulting risk. The designation of IROFS should generally be made to reduce the likelihood (i.e., prevent an accident), but the consequences may also be reduced by minimizing the potential hazards (e.g., quantity), if practical. Based on hazards identification and accident sequence analyses for which the resulting unmitigated or uncontrolled risks are unacceptable, key safety controls (administrative and/or engineered IROFS) may be designated as IROFS to reduce the likelihood of occurrence and/or mitigate the consequence severity.

The accident evaluation method described below does not preclude the need to comply with the double-contingency principle for sequences leading to criticality (see Chapter 5 of this SRP).

## A.5  Example Risk Index Evaluation Method

As previously mentioned, one acceptable way for the applicant to present the results of the ISA is a tabular summary of the identified accident sequences. Table A-7 shows an acceptable format for such a table. This table lists several example accident sequences for a powder blender at a typical facility. Table A-7 summarizes two sets of information: (1) the accident sequences identified in the ISA and (2) a risk index, calculated for each sequence, to show compliance with the regulation. This risk index is a representation of the frequency of the accident sequence in accordance with the mathematics underlying accidents resulting from sequences of events. The next section describes the underlying mathematics of this approach.

### A.5.1 Mathematics of Accident Sequence Frequencies and the Risk Index Method

According to 10 CFR 70.61, controls must be applied so that high-consequence events are highly unlikely and intermediate-consequence events are unlikely. This means that each accident sequence, consisting of initiating events and subsequent events, that leads to high consequences must be highly unlikely. In quantitative terms, "highly unlikely" will be treated here in terms of annual frequency of occurrence. The purpose of this section is to explain the concepts and mathematical formulas underlying the risk index method of likelihood evaluation, which is cited in this appendix as one example of an acceptable method for such evaluations in ISAs.

Since high-consequence events are, for workers, potentially life threatening or fatal, "highly unlikely" must be taken to mean of quite low frequency. Generally, achieving such low frequency requires either redundancy, robust passive control with large safety margin, or rare external events. Redundancy of safety controls is a method for limiting the occurrence rate of accidents by applying controls such that two coincident failure conditions must exist for a high-consequence event to occur. Use of redundant controls is common in criticality safety, where the double-contingency principle is standard. There are different types of redundant control systems. The effectiveness of each of these systems depends not just on having controls with low failure rates, but also on limiting downtime after failure occurs. Downtime, or the period of vulnerability resulting from an event, may be limited by the inherent fail-safe or failure-evident nature of the event. For events lacking these properties, failure should be detected either by hardware monitoring or by surveillance testing, which is usually part of the plant preventive maintenance program. Definition of the following symbols will aid in understanding how accident frequencies depend on frequency of failure events and downtime:

$\lambda_i$ = rate of failure of control i or of occurrence of initiating event i (in units of per year)
t = mean time to failure (MTTF) = $1/\lambda_i$ = mean uptime
$T_i$ = mean downtime of control i = $1/\mu_i$
$u_i$ = unavailability of control i
sfr = system failure rate (accident rate)

Mean downtime is often not the same as mean time to actually repair the affected safety system (MTTR), but rather indicates the mean time that the system is vulnerable to the second failure. This may be considerably shorter than the MTTR, if there is an alternative means of placing the system in a state as safe as it was with the unfailed control.

Unavailability, u, is defined as the probability that a control or system is not available to perform its function at a particular time. Unavailability is usually the predominant component of probability of failure of a system on demand. The normal model is that a control or system is either in an unavailable ("down") state or an available ("up") state. The system randomly changes from one state to the other over time, governed by the failure rate $\lambda$ and the repair rate $\mu = 1/T$. As a long-run average, the unavailability of a control is thus the fraction of the time that it is down, which is the ratio of downtime to downtime plus uptime:

$$u = T/(t+T)$$

For any reasonably available system, uptime is much greater than downtime, t >> T.

Thus, approximately, $u \approx T/t$ and $t = 1/\lambda$, so that $u \approx \lambda T$.

The three most common types of redundant control systems have the following equations for their system failure rate (accident rate):

two continuous parallel controls: $sfr = \lambda_1 u_2(1 - u_1) + \lambda_2 u_1(1 - u_2)$
usually approximated as: $sfr \approx \lambda_1 u_2 + \lambda_2 u_1 \approx \lambda_1(\lambda_2 T_2) + \lambda_2(\lambda_1 T_1)$

Equation (1)

three continuous parallel controls: $sfr = \lambda_1 u_2 u_3(1 - u_1) + \lambda_2 u_1 u_3(1 - u_2) + \lambda_3 u_1 u_2(1 - u_3)$
usually approximated as: $sfr \approx \lambda_1 u_2 u_3 + \lambda_2 u_1 u_3 + \lambda_3 u_1 u_2$

Equation (2)

challenging initiating event of frequency $\lambda_1$ with one control: $sfr = \lambda_1 u_2$

Equation (3)

initiating event i with two redundant standby identical controls: $sfr = \lambda_i u_1 u_2$

Equation (4)

The system of frequency and probability (of failure on demand) described in this appendix is based on taking the logarithm of each of the terms in the above equations. Thus, for Equation (1), in log space, two terms would correspond to the two accident sequences by which the system could fail, namely control 1 first or control 2 first:

sequence 1: $\log(\lambda_2) + \log(u_1)$
sequence 2: $\log(\lambda_1) + \log(u_2)$

If only failure rates $\lambda$ and downtimes T are used, then, with the approximation $u \approx \lambda T$, the formulas corresponding to Equation (1) become the following:

$sfr = \lambda_1(\lambda_2 T_2) + \lambda_2(\lambda_1 T_1)$

sequence 1: $\log(\lambda_2) + \log(\lambda_1) + \log(T_1)$
sequence 2: $\log(\lambda_1) + \log(\lambda_2) + \log(T_2)$

Thus, for two continuous redundant controls, two accident sequences are typically scored for likelihood. One of the two will usually have a larger frequency, so it is important to evaluate both. For situations modeled by Equation (3), there would be just one term.

Table A-9 below provides one example of criteria that might be used to assign frequency index numbers ($\log(frequency) = \log(\lambda)$). Table A-10 provides one example of criteria that might be used to assign index numbers for probabilities of failure on demand ($\log(unavailability) = \log(u)$). Table A-11 provides one example of criteria for assigning index numbers for downtime, that is, logarithms of durations of vulnerability, $\log(T)$. Note that when MTTF >> MTTR, $u = \lambda T$ approximately, so that the values $\lambda$ from Table A-9 and the values T from Table A-11 can be combined to obtain u for a given control if $\lambda$ and T are the known quantities.

The "average" downtime, when determined by surveillance, depends on the interval of time between scheduled system surveillance tests. If a surveillance test is done weekly, then, when the system is found to be in a failed state, the time that it could have been in this state is between zero and 1 week. Thus, the average time that the system will have been down, when discovered by the test, is half this, or 3.5 days. In units of per year, this is 3.5/365 = 0.01 year, and log(.01) = -2. Thus, a short surveillance interval can considerably reduce the system failure rate.

## A.5.2 An Example Application of a Risk Index Method of Likelihood Evaluation

Accident sequences result from initiating events, followed by failure of one or more IROFS. Thus, Table A-7 has columns for the initiating event and for IROFS. The initiating event may be failure of one of the IROFS, which may be mitigative or preventive. Mitigative IROFS are measures that reduce the consequences of an accident. In accordance with Tables A-9 through A-11, index numbers are assigned to initiating events, IROFS failure events, and mitigation failure events, based on the reliability characteristics of these items.

As an example, with two redundant IROFS, there is an accident sequence in which an initiating failure of one IROFS places the system in a vulnerable state. While the system is in this vulnerable state, the second IROFS may fail, which would result in an accident with consequences exceeding the criteria in 10 CFR 70.61. For such sequences, the frequency of the accident depends on three quantities: the frequency of the first event, the duration of vulnerability, and the frequency of the second IROFS failure. For this reason, the duration of the vulnerable state should be considered, and a duration index should be assigned. The values of all index numbers for a sequence are added to obtain a total likelihood index, T. In this risk index method of evaluation, accident sequences are then assigned to one of the three likelihood categories of the risk matrix, depending on the value of this index in accordance with Table A-8.

The values of index numbers in accident sequences are assigned considering the criteria in Tables A-9 through A-11. Each table applies to a different type of event. Table A-9 applies to events that have frequencies of occurrence, such as initiating events, which may be IROFS failures or external events. When failure probabilities are required for an event subsequent to the initiating event, Table A-10 provides the index values. Table A-11 provides index numbers for durations of failure. These are used in cases where information on probability of failure on demand is not available for the IROFS failures subsequent to the initiating event. Note the third row in Table A-7; it evaluates the reverse sequence to that in the first row. That is, the second IROFS fails first. This should be considered as a separate accident sequence, because, as shown, it may have a different frequency.

## Table A-7 Example Accident Sequence Summary and Risk Index Assignment

Process: uranium dioxide (UO₂) powder preparation (PP)
Unit Process: additive blending
Node: blender hopper node (PPB2)

| Accident Identifier A | Initiating Event or IROFS 1 Failure B | Preventive Safety Parameter 2 or IROFS 2 Failure/Success C | Mitigation IROFS Failure/ Success D | Likelihood* Index T E=B+C+D | Likelihood Category F | Consequence Category G | Risk Index H=F+G | Comments and Recommendations |
|---|---|---|---|---|---|---|---|---|
| **PPB2-1A** (Criticality from blender leak of UO₂) | PPB2-C1: Mass Control Failure: Blender leaks UO₂ onto floor; critical mass exceeded Frq1 = -1 Dur1 = -4 | PPB2-C2: Moderation Failure: Sufficient water for criticality introduced while UO₂ on floor: Frq2 = -2 | N/A | T = -7 | 1 | 3 | 4 | Criticality, consequences = 3, IROFS 2 fails while IROFS 1 is in failed state. T = -1-4-2 = -7 |
| **PPB2-1B** (Radiation release from blender leak of UO₂) | PPB2-C1: Mass Control Failure: Mass control fails but critical mass not exceeded Frq1=-1 Dur1 N/A | PPB2-C2: N/A | Ventilation Failure: Ventilated blender enclosure Prf = -3 | T = -4 | 1 | 2 | 3 | Rad consequences, no criticality unmitigated sequence: IROFS 1 and mitigation fail. T= -1-3 = -4 |
| **PPB2-1C** (Criticality from presence of water under blender) | PPB2-C2: Moderation Failure: Sufficient water for criticality on floor under UO₂ blender Frq1 = -2 Dur1 = -3 | PPB2-C1: Mass Control Failure: Blender leaks UO₂ on floor while water present Frq2 = -1 | N/A | T = -6 | 1 | 3 | 4 | Criticality by reverse sequence of PPB2-1A. Moderation fails first. Note different likelihood. T = -6 |

## Table A-8 Likelihood Category Assignment

| Likelihood Category | Likelihood Index T* (= sum of index numbers) |
|---|---|
| 1 | T ≤ -5 |
| 2 | -5 < T ≤ -4 |
| 3 | -4 < T |

## Table A-9  Failure Frequency Index Numbers

| Frequency Index No. | Based on Evidence | Based on Type of IROFS** | Comments |
|---|---|---|---|
| -6* | External event with frequency < $10^{-6}$/yr | | If initiating event, no IROFS needed. |
| -4* | No failures in 30 years for hundreds of similar IROFS in industry | Exceptionally robust passive engineered IROFS (PEC), or an inherently safe process, or two independent active engineered IROFS (AECs), PECs, or enhanced administrative IROFS | Rarely justified by evidence. Further, most types of single IROFS have been observed to fail. |
| -3* | No failures in 30 years for tens of similar IROFS in industry | A single IROFS with redundant parts, each a PEC or AEC | |
| -2* | No failure of this type in this facility in 30 years | A single PEC | |
| -1 | A few failures may occur during facility lifetime | A single AEC, an enhanced administrative IROFS, an administrative IROFS with large margin, or a redundant administrative IROFS | |
| 0 | Failures occur every 1 to 3 years | A single administrative IROFS | |
| 1 | Several occurrences per year | Frequent event, inadequate IROFS | Not for IROFS, just initiating events. |
| 2 | Occurs every week or more often | Very frequent event, inadequate IROFS | Not for IROFS, just initiating events. |

\*  Indices less than (more negative than) -1 should not be assigned to IROFS unless the configuration management, auditing, and other management measures are of high quality, because without these measures, the IROFS may be changed or not maintained.

\*\*  Failure frequencies based on experience for a particular type of IROFS, as described in this column, may differ from values in column 1; in this case, data from experience take precedence.

## Table A-10  Failure Probability Index Numbers

| Probability Index No. | Probability of Failure on Demand | Based on Type of IROFS | Comments |
|---|---|---|---|
| -6* | $10^{-6}$ | | If initiating event, no IROFS needed. |
| -4 or -5* | $10^{-4}$–$10^{-5}$ | Exceptionally robust passive engineered IROFS (PEC), or an inherently safe process, or two redundant IROFS more robust than simple administrative IROFS (AEC, PEC, or enhanced administrative) | Rarely justified by evidence. Most types of single IROFS have been observed to fail. |
| -3 or -4* | $10^{-3}$–$10^{-4}$ | A single passive engineered IROFS (PEC) or an active engineered IROFS (AEC) with high availability | |
| -2 or -3* | $10^{-2}$–$10^{-3}$ | A single active engineered IROFS, or an enhanced administrative IROFS, or an administrative IROFS for routine planned operations | |
| -1 or -2 | $10^{-1}$–$10^{-2}$ | An administrative IROFS that must be performed in response to a rare unplanned demand | |

\*    Indices less than (more negative than) -1 should not be assigned to IROFS unless the configuration management, auditing, and other management measures are of high quality, because without these measures, the IROFS may be changed or not maintained.

## Table A-11  Failure Duration Index Numbers

| Duration Index No. | Average Failure Duration | Duration in Years | Comments |
|---|---|---|---|
| 1 | More than 3 years | 10 | |
| 0 | 1 year | 1 | |
| -1 | 1 month | 0.1 | Formal monitoring to justify indices less than -1 |
| -2 | A few days | 0.01 | |
| -3 | 8 hours | 0.001 | |
| -4 | 1 hour | $10^{-4}$ | |
| -5 | 5 minutes | $10^{-5}$ | |

As shown in Table A-11, the duration of failure, and thus the period that the system is in a state of heightened vulnerability, is accounted for in establishing the overall frequency of the accident sequence. The period of vulnerability will normally be terminated by discovery of the vulnerable condition or failure; the system will then be rendered safe, either by removing the hazardous material, or by repairing or substituting for the safety function of the failed IROFS. The duration of this period of vulnerability determines the index value to be assigned from Table A-11.

For all these index numbers, the more negative the number, the lower the frequency of the event. Accident sequences may consist of varying numbers of events, starting with an initiating event. The total likelihood index is the sum of the indices for all the events in the sequence, including those for duration, except the initiating event, for which only the occurrence frequency index should be used. For example, a three-event sequence would correspond to an event sequence frequency of the form $\lambda_1(\lambda_2 T_2)(\lambda_3 T_3)$, or five index values, three being frequencies, and two durations.

Consequences are assigned to one of the three consequence categories of the risk matrix, based on calculations or estimates of the actual consequences of the accident sequence. The consequence categories are based on the levels identified in 10 CFR 70.61. Multiple types of consequences can result from the same event. If there are multiple types of consequence, the consequence category is the most severe. Similarly, if a range of consequences could occur, then the highest consequence event of this range could occur, and if it falls in the high-consequence range, it should be evaluated as such.

Table A-12 provides a more detailed description of the accident sequences used in the example of Table A-7. Such descriptive information may be necessary for the reviewer to understand the nature of the accident sequences listed in Table A-7.

Table A-13 is an example of one format for the descriptive list of IROFS required by the regulation. It should also include external initiating events that appear in the accident sequences and whose frequencies are relied on in demonstrating that the overall accident sequence frequency complies with the likelihood requirements. The information on IROFS in Table A-13 should have sufficient detail to permit the reviewer to understand why the initiating events and IROFS listed in Table A-7 have the frequency, unavailability, or duration indices assigned. Thus, Table A-13 may also contain such information as (1) the margins to safety limits, (2) the redundancy of an IROFS, and (3) the measures taken to ensure adequate reliability of an IROFS, if this information is necessary to understand the reliability and safety function of the IROFS with respect to the likelihood performance requirements.

**Table A-12  Accident Sequence Descriptions**

Process:  UO$_2$ powder preparation (PP)
Unit:  additive blending
Node:  blender hopper node (PPB2)

| Accident (see Table A-6) | Description |
|---|---|
| PPB2-1A Blender UO$_2$ leak criticality | The initial failure is a blender leak of UO$_2$ that results in a mass sufficient for criticality on the floor. (This event is not a small leak.) Before the UO$_2$ can be removed, moderator sufficient to cause criticality is introduced. Duration of critical mass UO$_2$ on floor is estimated to be 1 hour. |
| PPB2-1B Blender UO$_2$ leak, radiological release | The initial failure is a blender leak of UO$_2$ that results in a mass insufficient for criticality on the floor or a mass sufficient for criticality but moderation failure does not occur. Consequences are radiological, not a criticality. A ventilated enclosure should mitigate the radiological release of UO$_2$. If it fails during cleanup or is not working, unmitigated consequences occur. |
| PPB2-1C | The events of PPB2-1A occur in reverse sequence. The initial failure is introduction of water onto the floor under the blender. Duration of this flooded condition is 8 hours. During this time, the blender leaks a critical mass of UO$_2$ onto the floor. Criticality occurs. |

**Table A-13  Descriptive List of IROFS**

Process:  UO$_2$ powder preparation (PP)
Unit:  additive blending
Node:  blender hopper node (PPB2)

| IROFS Identifier | Safety Parameter and Limits | IROFS Description | Max. Value of Other Parameters | Reliability Management Measures | Quality Assurance Grade |
|---|---|---|---|---|---|
| PPB2-C1 | Mass outside hopper: zero | Mass outside hopper: Hopper and outlet design prevent UO$_2$ leaks, double gasket at outlet | Full water reflection, enrichment 5% | Surveillance for leaked UO$_2$ each shift | A |
| PPB2-C2 | Moderation: in UO$_2$ < 1.5 wt.% External water in area: zero | Moderation in UO$_2$: Two sample measurements by two persons before transfer to hopper External water: Posting excluding water, double piping in room, floor drains, roof integrity | Full water reflection, enrichment 5% | Drain, roof, and piping under safety-grade maintenance | A |

Note:    In addition to IROFS, which are facility hardware and procedures, this table should describe external initiating events, the low likelihood of which is relied on to achieve acceptable risk, especially those that are assigned frequency indices lower than -4. The descriptions of these initiating events should contain information supporting the frequency index value selected by the applicant.

## A.6  Determination of Likelihood Category in Table A-8

The likelihood category is determined by calculating the likelihood index, T, which equals the sum of the indices for the events in the accident sequence. Based on the calculated value of T, the likelihood category of each accident sequence can be determined from Table A-8.

## A.7  Failure Probability Index Numbers in Table A-10

Occasionally, information concerning the reliability of an IROFS may be available as a probability on demand. That is, there may be a history of tests or incidents where the system in question is demanded to function. To quantify such accident sequences, the demand frequency, the initiating event, and the demand failure probability of the IROFS must be known. This table provides an assignment of index numbers for such IROFS in a way that is consistent with Table A-9. The probability of failure on demand may be the likelihood that it is in a failed state when demanded (availability) or that it fails to remain functional for a sufficient time to complete its function.

## A.8  Management Measures for Items Relied on for Safety

Table A-13 is an acceptable way of listing IROFS in all the general types of accident sequences having consequences exceeding those identified in 10 CFR 70.61. The items listed should include all IROFS and all external events whose low likelihood of occurrence is relied on to meet the performance requirements of 10 CFR 70.61. For certain IROFS or accident sequences included in this list, information on management measures that is specific to that sequence or IROFS should be presented to permit the reviewer to understand how the IROFS perform. The reviewer examines this list to determine whether adequate management measures have been applied to each IROFS to ensure its continual availability and reliability, in conformance with 10 CFR 70.62(d). Management measures include such activities as maintenance, training, configuration management, audits and assessments, and quality assurance. The baseline design criteria indicate criteria for management measures; SRP Chapters 4 through 7 and Chapter 11 describe other criteria in greater detail. IROFS may have management measures applied in varying ways or to varying degrees, depending on the nature of the IROFS, and the degree of reliability assumed in demonstrating compliance with the likelihood requirements. This is the meaning of "graded management measures."

## A.9  Risk-Informed Review of Items Relied on for Safety

Column (H) in Table A-7 gives the risk indices for each accident sequence identified in the ISA. There are two indices, uncontrolled and controlled. The controlled index is a measure of risk without credit for the IROFS. If the uncontrolled risk index is a 6 or 9, while the controlled index is an acceptable value (4 or less), the set of IROFS involved are significant in achieving acceptable risk. That is, these IROFS have high risk significance. Reviewers will use the uncontrolled risk index to identify all risk-significant systems of IROFS. These systems of IROFS will be reviewed more closely than IROFS established to prevent or mitigate accident sequences of low risk.

## REFERENCES

*U.S. Code of Federal Regulations*, Chapter I, Title 10, "Energy," Part 20, "Standards for Protection Against Radiation."

*U.S. Code of Federal Regulations*, Chapter I, Title 10, "Energy," Part 70, "Domestic Licensing of Special Nuclear Material."

U.S. Nuclear Regulatory Commission, "Nuclear Fuel Cycle Facility Accident Analysis Handbook," NUREG/CR-6410, March 1998.

# ANNEX TO APPENDIX A

# USE OF APPENDIX A RISK INDEX METHODOLOGY

## Introduction

The purpose of this annex is to clarify the proper use of the semiquantitative index method as described in Appendix A to Chapter 3 of this Standard Review Plan. Several licensees and applicants have used the index method of Appendix A (or a variation thereof) in performing their integrated safety analyses (ISAs). The U.S. Nuclear Regulatory Commission (NRC) reviews of these ISA Summaries have discovered a need for additional guidance on the use of this method. Because of the method's widespread use and a lack of common understanding about its use, guidance on the index method is appropriate.

As stated in the introduction to Appendix A, the index method is but one method of likelihood evaluation. The index method is not strictly a qualitative method; rather, it is a semiquantitative method that considers both qualitative and quantitative information (if it is available and applicable).. In this method, the definition of likelihood terms (i.e., "not unlikely," "unlikely," and "highly unlikely") is expressed quantitatively (more than $10^{-4}$ per event, per year; between $10^{-4}$ and $10^{-5}$ per event, per year; and less than $10^{-5}$ per event, per year, respectively). As a purely qualitative method would use purely qualitative definitions of likelihood and qualitative methods of evaluating likelihood, much of the quantitative discussion in this appendix would not apply. However, this method illustrates the logic that should be used in even a purely qualitative method.

The index method is one acceptable method of demonstrating compliance with the performance requirements. However, taking credit for using this method requires that the applicant follow all of the guidance contained in Appendix A. Otherwise, the applicant should provide additional justification.

## Likelihood Definitions

The likelihood definitions in Table A-6 of Appendix A are, as stated above, given in quantitative terms (e.g., "highly unlikely" is defined as less than $10^{-5}$ per event, per year). The footnote to Table A-6 indicates, however, that these are based on approximate order-of-magnitude ranges. Therefore, these values should not be regarded as strict numerical limits but as indicative of the approximate order of magnitude of likelihood. Any definition of likelihood should be stated on a per-event basis.

## Likelihood Evaluation Method

The likelihood evaluation method used should be consistent with the likelihood definitions, such that the qualitative score assigned can be compared to the likelihood definitions. In the index method, the likelihood index for the accident sequence must be no greater than -5 to meet the definition of "highly unlikely" and must be no greater than -4 to meet the definition of "unlikely." The likelihood index for the accident sequence is determined by summing likelihood indices for the initiating event and subsequent failures of items relied on for safety (IROFS). Tables A-9, A-10, and A-11 of Appendix A present criteria for the assignment of the likelihood indices.

Appendix A distinguishes between two different kinds of events that can be combined to form the accident sequences in the ISA Summary. The two basic kinds of events are (1) events that are characterized by a frequency of occurrence, and (2) events that are characterized by a probability of failure on demand (PFOD). In the index method of Appendix A, the category to which an event belongs determines how it is scored by means of either Table A-9 or A-10, as explained below.

Events characterized by a frequency of occurrence (f-type events) can include external events, internal events that are not IROFS failures, or IROFS failures. IROFS failures characterized by a frequency of occurrence are those that are required to be continuously present, rather than those that are required to perform a safety function only when certain conditions are present. Examples may include favorable geometry equipment or an active engineered device monitoring a continuous process.

Events characterized by a PFOD (p-type events) typically include IROFS that are not required to be continuously present but that must perform a safety function on demand (subsequent to some process deviation or failure). Examples include active interlocks that perform some protective function when system parameters exceed preset limits, administrative controls required in response to process deviations, or certain administrative controls in batch processes. These are usually part of the subsequent failures following the initiating event but may sometimes be part of the initiating event.

In general, accident sequences may comprise many individual events. In general, accident sequences consist of an initiating event followed by the failure of one or more IROFS. Because the overall accident sequence likelihood must be consistent with the likelihood categories, it must have the same dimensional units as those of the likelihood definitions (i.e., probability per event, per year). Even though qualitative indices are used instead of quantitative probabilities, this requirement imposes constraints on the ways in which individual indices may be combined.

For simplicity, the following considers only two-event sequences (in which the events are independent). The two basic kinds of events result in four basic types of two-event accident sequences, as described in the following sections.

### F-Type Initiating Event with Subsequent P-Type IROFS Failure

In the index method of Appendix A, a failure frequency index may be applied to the initiating event using the criteria in Table A-9, and a failure probability index may be applied to the subsequent IROFS failure using the criteria in Table A-10. The overall likelihood index for the accident sequence is the sum of the likelihood indices for the two events. This is because the IROFS is assumed to be demanded every time the initiating event occurs.

Mathematically, this results in an accident sequence likelihood index corresponding to an accident sequence likelihood with the correct dimensional units:

- accident sequence likelihood ($yr^{-1}$) = initiating event frequency ($yr^{-1}$) × PFOD
  accident sequence index = initiating event index + subsequent failure index

An example of this type of accident sequence is a criticality sequence consisting of a loss of concentration control in a continuous solution processing operation, followed by failure of an inline concentration monitor that closes an isolation valve on a transfer line upon detection of highly concentrated solution.

### F-Type Initiating Event with Subsequent F-Type IROFS Failure

Using the index method of Appendix A, a failure frequency index may be applied to both the initiating event and the subsequent IROFS failure using the criteria in Table A-9. The overall likelihood index for the accident sequence is the sum of the individual likelihood indices for the two events and a duration index for the initiating event. This is because the probability of the second event occurring concurrently with the first event is dependent on the time during which the conditions caused by the first event persist. For the accident sequence likelihood to have the correct units ($yr^{-1}$), the duration of failure for the first event must be considered.

Mathematically, this results in an accident sequence likelihood index corresponding to an accident sequence likelihood with the correct dimensional units:

- accident sequence likelihood ($yr^{-1}$) = initiating event frequency ($yr^{-1}$) × initiating event duration (yr) × subsequent failure frequency ($yr^{-1}$)

- accident sequence index = initiating event index + initiating event duration index + subsequent failure index

An example of this type of accident sequence is a criticality sequence consisting of a loss of geometry control followed by a loss of moderation control resulting from the unrelated sprinkler activation before geometry control can be restored.

### P-Type Initiating Event with Subsequent P-Type IROFS Failure

Using the index method of Appendix A, a failure probability index may be applied to both the initiating event and the subsequent IROFS failure using the criteria in Table A-10. The overall likelihood index for the accident sequence is the sum of the individual likelihood indices for the two events, which includes consideration of the demand rate associated with the initiating event. This is because the total failure frequency for the initiating event depends on the frequency with which the demand occurs, as well as the associated PFOD. The subsequent IROFS is assumed to be demanded every time the initiating event occurs. For the accident sequence likelihood to have the correct units ($yr^{-1}$), the demand rate of the first event must be considered.

Mathematically, this results in an accident sequence likelihood index corresponding to an accident sequence likelihood with the correct dimensional units:

- accident sequence likelihood ($yr^{-1}$) = initiating event demand rate ($yr^{-1}$) × initiating event PFOD × subsequent event PFOD

- accident sequence index = initiating event index (including demand rate) + subsequent failure index

An example of this type of accident sequence is a criticality sequence consisting of the failure of an operator to sample solution before transfer in a batch operation, followed by failure of an inline concentration monitor as discussed previously.

**P-Type Initiating Event with Subsequent F-Type IROFS Failure**

Using the index method of Appendix A, a failure probability index may be applied to the initiating event using the criteria in Table A-10. A failure frequency index may be applied to the subsequent IROFS failure using the criteria in Table A-9. The overall likelihood index for the accident sequence is the sum of likelihood indices for the two events, which includes consideration of the demand rate associated with the initiating event and a duration index for the initiating event. This is because the failure frequency for the initiating event depends on the frequency with which the demand occurs, as well as the associated PFOD. The probability of the second event occurring concurrently with the first event is dependent on the time during which the conditions caused by the first event persist. For the accident sequence likelihood to have the correct units ($yr^{-1}$), both the duration of failure for the first event and its demand rate must be considered.

Mathematically, this results in an accident sequence likelihood index corresponding to an accident sequence likelihood with the correct dimensional units:

- accident sequence likelihood ($yr^{-1}$) = initiating event demand rate ($yr^{-1}$) × initiating event PFOD × initiating event duration (yr) × subsequent failure frequency ($yr^{-1}$)

- accident sequence index = initiating event index (including demand rate) + failure duration index + subsequent failure index

An example of this type of accident sequence is a criticality sequence consisting of a uranium solution spill that results from improper preventive maintenance on a pump, followed by the loss of moderation control because of inadvertent sprinkler activation before the spill can be cleaned up.

## Use of Tables A-9, A-10, and A-11 in Appendix A

As illustrated above, an accident sequence generally consists of an initiating event with a certain frequency, followed by a number of subsequent events. While the number and type of events making up the sequence may vary, the likelihood indices of the individual events are combined, with appropriate consideration for duration of failure and demand rate, to arrive at a likelihood index for the accident sequence as a whole. The basic steps in this process are outlined below:

(1)     Determine the events making up the sequence (initiating event and subsequent failures).

(2)     Determine whether the event is characterized by a frequency of occurrence (f-type) or a PFOD (p-type). If an f-type event, use Table A-9 to assign the indices. If a p-type event, use Table A-10 to assign the indices.

(3)     If the initiating event is a p-type event, take the demand rate into account to modify the indices from Table A-9.

(4)     If the subsequent event is an f-type event, take into account the duration index for the initiating event from Table A-11.

(5)     Combine the appropriate indices into an overall accident sequence likelihood index.

The table below summarizes the use of Tables A-9, A-10, and A-11 to determine overall accident sequence likelihood:

| Initiator Type | Subsequent Event Type | Initiator Index | Subsequent Event Index | Duration Index | Accident Sequence Index |
|---|---|---|---|---|---|
| f-type | p-type | f1: Table A-9 | p2: Table A-10 | NA | f1 × p2 |
| f-type | f-type | f1: Table A-9 | f2: Table A-9 | d1: Table A-11 | f1 × d1 × f2 |
| p-type | p-type | p1: Table A-10* | p2: Table A-10 | NA | p1 × p2 |
| p-type | f-type | p1: Table A-10* | f2: Table A-9 | d1: Table A-11 | p1 × d1 × f2 |

\*      To convert PFOD indices to frequency indices, use the indices of Table A-10 modified to take demand rate into account as follows:

| Demand Rate | Modify Table A-10 Index |
|---|---|
| Hundreds of times per year (daily) | Increase base index by 2 |
| Tens of times per year (monthly) | Increase base index by 1 |
| Once per year | Use base index |
| Once every 10 years | Decrease base index by 1 |

Users of these tables must be careful not to confuse frequency with probability. For example, it is often assumed that the initiating event occurs because soothe assumption is simpler and more conservative. This is not, however, equivalent to assigning an initiating event frequency of 1, which is an event that occurs once per year. The confusion of failure frequency (with units of inverse time) with probability (dimensionless) can lead to significant errors in the overall accident sequence likelihood.

Example: In this accident sequence, the initiating event is solution sampling before transfer to a tank with an unfavorable geometry. A single administrative control might have a probability index of -2 (with appropriate management measures or redundancy). Similarly, if the historical data indicated a PFOD of $10^{-2}$, an index of -2 would be appropriate. However, if this operation is a batch process conducted 10 times per year, this results in an initiating event frequency of 10/yr × $10^{-2}$ (PFOD) = $10^{-1}$/yr (for an index of -1). If the operation is conducted 100 times per year, this results in an initiating event frequency of 100/yr × $10^{-2}$ (PFOD) = $10^{0}$/yr (for an index of 0). Use of Table A-10 without any consideration of the demand rate would result in an index of -2.

Use of the incorrect table can also lead to erroneous results. A comparison of the indices in Tables A-9 and A-10 for the same type of control (although this is not the only factor that should be considered) immediately shows that use of Table A-9 results in a higher index than does use of Table A-10. For example, a simple administrative control (without enhancing factors such as redundancy or large margin) would have a probability index of -1 to -2 based on Table A-10, but a frequency index of 0 based on Table A-9. This is intuitively reasonable because Table A-9 is for events characterized by a frequency (which must be present on a continuous basis) and Table A-10 is for events that are demanded only under certain conditions (which must be present on occasion).

## Additional Considerations in the Use of Index Tables

Assignment of a qualitative score may be based either on objective evidence of the frequency of occurrence or on certain qualitative characteristics of the process or facility (availability and reliability qualities). In accordance with this, Tables A-9 and A-10 contain two columns that represent two different methods for assigning likelihood indices. As stated in the introduction to Appendix A, this is a semiquantitative method that allows for the use of quantitative information if available.

For initiating events that are external events or internal events other than IROFS failures, the column entitled "Based on Evidence" in Table A-9 should be used in assigning indices. For IROFS failures to which Table A-9 applies, either the column entitled "Based on Evidence" or "Based on Type of IROFS" may be used. Because the type of IROFS is only one of the availability and reliability qualities on which likelihood depends, the footnote to this table indicates that the index scores applicable to a particular type of IROFS can be one value higher or lower than the index shown.[1] Thus, other specific availability and reliability qualities (as discussed in Section 3.4.3.2(9) of this SRP) should be considered in assigning the final likelihood index.[2] In the absence of sufficiently detailed information about these factors, appropriate conservatism should be used in assigning indices (e.g., using the highest index in the range). Because of the large uncertainty associated with basing likelihood on the type of IROFS, historical and/or operating evidence should be used to assign indices whenever available. The same considerations discussed above should be employed when using Table A-10 to assign likelihood indices.

The presence of two columns should not be construed to mean that the two sets of criteria may be considered equivalent except in a rough, order-of-magnitude sense (e.g., a single passive engineered IROFS does not necessarily have a PFOD of $10^{-3}$ to $10^{-4}$). This is because the type of IROFS is only one of the availability and reliability qualities that must be considered.

Appropriate use of Tables A-9 and A-10 to assign likelihood indices also requires that attention be given to the footnotes and comments in these tables. As indicated in the footnotes, indices less than -1 should not be used unless the management measures are of high quality. This is because even though a passive engineered control may have high inherent reliability while it is installed, this control could be easily defeated by a poor configuration management program, which is administrative in nature (as are all management measures). Justification should be

---

[1]    The title "Based on Type of IROFS" is somewhat of a misnomer in that several of the criteria also include consideration of redundancy, margin, and independence. Indices based solely on the type of IROFS would cover an even broader range.

[2]    This is consistent with the caveat for Table A-4, which warns that such coarse criteria are useful only for screening purposes or making an initial estimate of the likelihood. Because IROFS meeting these criteria can have a broad range of reliability, management measures applied to all the availability and reliability qualities of the IROFS should be considered in assigning the likelihood indices.

provided as to why the management measures are deemed to be of high quality. Also, the ISA Summary should justify the use of a more negative index whenever a range of indices is possible. As the comments suggest, the more negative the index, the more justification is required. As indicated, indices of -4 and -5 can rarely be justified by evidence. Use of these indices requires substantial evidence that the IROFS are exceptionally robust.

The assignment of failure duration indices using Table A-11 should also be based on objective criteria (such as documented mean time to repair or surveillance periods established in plant procedures).

When the analysis uses demand rates to modify probability indices from Table A-10, conservative estimates of the demand rate should be used and the basis for this estimate documented and, if the rates could credibly be changed, controlled. For example, the time needed to fill a cylinder may depend on inherent physical laws and would not need specific controls. However, if the maximum allowed inventory limits the number of batches, the license or plant procedures should control this inventory.

**Description of Accident Sequences and IROFS**

Tables A-12 and A-13 include descriptions of accident sequences and IROFS. These must be sufficiently clear to permit the reviewer to understand the sequence of events needed for an accident to occur and how the established controls prevent the sequence from occurring. The initial failure and all subsequent failures necessary for the sequence to progress to the ultimate consequences (an accident exceeding the consequence thresholds in Title 10 of the *Code of Federal Regulations* (10 CFR) 70.61, "Performance Requirements") should be specified. In addition, any initial conditions credited in meeting the performance requirements should be specified. If important to the likelihood of the sequence, the order in which these events occur should be specified. For example, in Table A-12, sequence PPB2-1C is the reverse of the events in sequence PPB2-1A. When failure duration indices are considered, these pertain to the initiating event; therefore, the accident sequence likelihood is dependent on which event occurs first.

In describing IROFS, it is important that the safety function performed by the IROFS and the attributes of the IROFS necessary to perform the safety function be specified. For example, for the first IROFS in Table A-13, the safety function is to prevent mass from accumulating outside the hopper. Therefore, the only attribute of IROFS PPB2-C1 that must be specified is that it be designed to prevent leaks; such a design would include the use of a double gasket at the hopper's outlet. Because the material of composition, size, and other attributes of the hopper have no role in preventing this accident sequence, they need not be specified. The second IROFS is an example of a system of IROFS that collectively provides for moderation control (i.e., dual sampling, administrative exclusion of water, double piping, floor drains, and roof integrity). As in the preceding example, the size of the piping is not significant; double piping is the only feature important to preventing this accident sequence. The level of detail should be sufficient to provide assurance that safety-significant aspects of the IROFS are recognized and appropriately controlled. However, excessive detail could lead to obscuring the safety-significant aspects of IROFS and could lead to unnecessary and burdensome changes to the ISA and ISA Summary. IROFS may be specified at the subcomponent level, component level, or system level, as appropriate. For example, it is not necessary to specify every geometry limited pipe in the building as an IROFS. If the safety function is to maintain geometry control, it would be sufficient to specify a systems-level IROFS with the description "all fissile material piping in the solution recovery area will be less than 2 inches in diameter."

A single piece of equipment may perform several different safety functions and be credited in several different accident sequences. In such cases, the accident sequence must clearly describe the safety function and attribute of the IROFS being credited, as well as the failure mode of the IROFS that leads to the accident.

## Summary Table of Accident Sequences

Table A-7 of Appendix A is a summary table showing several accident sequences for a powder-blending process. This is one way to display the information on accident sequences obtained during performance of the ISA. As shown in Appendix A to NUREG-1718, "Standard Review Plan for the Review of an Application for a Mixed Oxide Fuel Fabrication Facility," issued August 2000, a fault tree (quantitative or qualitative) is one of the other formats that may be used. The important information that must be conveyed, however, is a list of accident sequences, identification of the initiating event, the set of subsequent events leading to the accident and the IROFS that prevent them, the likelihood of the initiating event and subsequent failures, the ultimate consequence category, and the overall assessment of compliance with the performance requirements (e.g., total risk index). Any other information needed to demonstrate that the performance requirements are met should also be specified (e.g., initial conditions, demand rate, duration indices, index modification for dependent failures). Table A-7 shows two types of accident sequences: (1) two sequences initiated by IROFS failures (both f-type initiating events with f-type subsequent failures and crediting duration indices) and (2) two sequences initiated by internal events other than IROFS failures (and crediting initiating event frequency).

While this guidance follows the structure of Appendix A to this Standard Review Plan, it also applies to Appendix A to NUREG-1718.

**REFERENCES**

*U.S. Code of Federal Regulations*, Chapter I, Title 10, "Energy," Part 70, "Domestic Licensing of Special Nuclear Material."

U.S. Nuclear Regulatory Commission, "Standard Review Plan for the Review of an Application for a Mixed Oxide (MOX) Fuel Fabrication Facility," NUREG-1718, August 2000.

# APPENDIX B

# QUALITATIVE CRITERIA FOR EVALUATION OF LIKELIHOOD

## Purpose

This appendix provides additional guidance on the use of qualitative criteria in methods for evaluation of likelihood. These evaluations are used in demonstrating compliance with the performance requirements of Title 10 of the *Code of Federal Regulations* (10 CFR) 70.61, "Performance Requirements."

## Introduction

The regulation in 10 CFR 70.61(b) requires that the risk of each credible high-consequence event be limited by ensuring that upon implementation of engineered or administrative controls, the event is made highly unlikely or its consequences reduced to less than high consequence. This regulation similarly requires that the risk of each credible intermediate-consequence event be limited by ensuring that the event is made unlikely, or its consequences reduced. Rather than defining the terms "highly unlikely," "unlikely," and, "credible," 10 CFR 70.65(b)(9) instead states that the applicant must include definitions of these terms in its integrated safety analysis (ISA) summary.

As stated in Section 3.4.3.2(9) of this Standard Review Plan (SRP), the applicant's definitions of these terms may be either quantitative or qualitative. The method used to evaluate accident sequence likelihood must be consistent with the definitions. Quantitative definitions require quantitative methods; qualitative definitions require qualitative methods. Qualitative methods are based on objective qualitative criteria and characteristics of the process or system being evaluated. In addition, some methods (semiquantitative methods) may rely on a mixture of qualitative and quantitative definitions, methods, and information. This appendix provides general guidance on the use of qualitative methods for evaluation of likelihood. However, the U.S. Nuclear Regulatory Commission's (NRC's) review of recently submitted ISA Summaries has revealed a lack of common understanding as to what constitutes an acceptable qualitative method.

Additional guidance is provided on the acceptance criteria for qualitative methods of evaluating likelihood, both for the failure of items relied on for safety (IROFS) and for accident sequences as a whole. Either external events or internal events (which may or may not be IROFS failures) may initiate these accident sequences. Appendix D to Chapter 3 of this SRP provides additional guidance on the use of initiating events that are natural phenomena. Appendix C to Chapter 3 offers additional guidance on the use of initiating events that are internal to the facility. That guidance may be used with the guidance in this appendix as an acceptable qualitative method for likelihood evaluation.

## Discussion

### Definitions of Likelihood

According to 10 CFR 70.65(b)(9), the ISA Summary must define the terms "unlikely," "highly unlikely," and "credible." Section 3.4.3.2(9) of this SRP states that qualitative definitions of likelihood are acceptable if they meet two conditions: (1) they are reasonably clear and based on objective criteria and (2) they can reasonably be expected to consistently distinguish accidents that are highly unlikely from those that are merely unlikely (or not unlikely). This means that the definitions should be sufficiently clear that there is reasonable assurance that they will yield the same result when applied by different reviewers and that they can be used to make meaningful distinctions between events in different likelihood categories. Both the definitions of likelihood and the methods for likelihood determination should meet these criteria since they must work together to ensure that the performance requirements are met.

This NUREG states that "objective criteria" means that the method relies on specific identifiable characteristics of a process design, rather than subjective judgments of adequacy. Because the likelihood of an accident sequence is a function of the likelihood of the initiating event, the subsequent IROFS failures, and the relationship between the IROFS (e.g., whether the IROFS are independent), the characteristics of the process design that the method should rely on are the specific identifiable characteristics of the initiating event, IROFS failures, and other process features that affect the likelihood of the accident sequence. These features include the safety margin, type of control, type and grading of management measures, whether the system is fail-safe or failure is self-announcing, failure modes, demand rates, and failure rates for individual IROFS (whether credited as part of the initiating event or subsequent failures). These features include the degree of redundancy, independence, diversity, and vulnerability to common-cause failure for systems of IROFS. The following sections describe these features in detail. It is important that any features of the process or equipment necessary to meet the performance requirements are recognized as important to safety and appropriately maintained through the use of management measures.

Examples of acceptable qualitative definitions of likelihood are the second and third definitions of "not credible" in Section 3.4.3.2(9) of this SRP:

> A process deviation consists of a sequence of many unlikely human actions or errors for which there is no reason or motive....

> A convincing argument exists that, given physical laws, the process deviations are not possible, or are unquestionably extremely unlikely....

Similarly, the following is an example of an acceptable qualitative definition of "highly unlikely":

> a system of IROFS that possesses double-contingency protection, where each of the applicable qualities is present to an appropriate degree

In this definition, the qualities to be considered should be described in sufficient detail so that their effect on the overall likelihood can be evaluated. This is the meaning of "present to an appropriate degree." Other definitions are acceptable provided that they meet the two criteria specified above and provide system features to ensure that the likelihood is appropriately maintained.

## Evaluation of Likelihood

Accident sequences, in general, consist of an initiating event followed by one or more subsequent events. The likelihood of an accident sequence is, therefore, a function of the likelihood of the individual events making up the accident sequence and the relationship between them (e.g., whether they are independent). Because the likelihood of the accident sequence must be compared to the likelihood definitions to determine whether it is "unlikely," "highly unlikely," or "not unlikely," qualitative methods of likelihood evaluation are acceptable if they (1) are reasonably clear and based on objective criteria and (2) can reasonably be expected to consistently distinguish accidents that are "highly unlikely" from those that are merely "unlikely." The likelihood definitions establish the standard for what is "unlikely" and "highly unlikely," and the assigned likelihood for the accident sequence is then compared to this standard. As mentioned above, the method must take into account all objective qualities of the system that can reasonably be considered to affect likelihood. These qualities are referred to in this NUREG as the "reliability and availability" qualities of IROFS or systems of IROFS.

## Initiating Events and Initial Conditions

Each accident sequence begins with an initiating event. An initiating event may consist of an external event (including a natural phenomenon or external manmade event), an internal event other than an IROFS failure, or an IROFS failure. Natural phenomena may include heavy rains, winds, flooding, earthquakes, and fires. External manmade events may include impacts from nearby facilities, aircraft or vehicle crashes, fires, and loss of offsite utilities. Internal events other than IROFS failures may include spills, non-IROFS equipment failure, process deviations, industrial accidents, and loss of onsite utilities. In a qualitative method of likelihood determination, a qualitative score is associated with the initiating event based on its objective qualities. The score may be expressed in either numerical (e.g., -1, -2, -3) or nonnumerical (e.g., A, B, C, D) form but is still qualitative if based on qualitative criteria.

The likelihood of external initiating events (by definition, they are outside the control of the facility) does not rely on any design features of the facility or process and is thus characterized only by a frequency of occurrence. In a qualitative method for assigning likelihood to these events, a qualitative score is associated with the external event based on its frequency of occurrence. Events with the same frequency of occurrence should have the same score regardless of the type of event or severity of its consequences. The method should thus include a table of the scores assigned based on qualitative frequency criteria. These criteria may include qualitative descriptions of frequency, such as "100-year flood" or "1,000-year earthquake," or may include other qualitative criteria that can be correlated to a frequency, such as "design-basis earthquake" or "exceeds the mean annual rainfall by a factor of x." By contrast, quantitative or semiquantitative methods may include quantitative descriptions of frequency such as "having a frequency less than $10^{-2}$ per year." Because these events are beyond human control, no features have to be maintained to ensure the continued validity of the assigned likelihood. However, it may be necessary to periodically reexamine the basis of these likelihoods if it is reasonably expected that the likelihood could change (e.g., following construction of a new railroad spur next to the facility). Appendix D to Chapter 3 contains additional guidance applicable to initiating events that are natural phenomena.

By contrast, the likelihood of internal initiating events other than IROFS failures depends on specific, identifiable characteristics of the facility or process design, such as those discussed in the following sections. Scores may be assigned to such events based either on objective evidence of their frequency of occurrence or on specific identifiable characteristics of the facility or process that can affect the frequency of occurrence. If the actual frequency of occurrence is known, this information should be used as it represents objective knowledge about the event likelihood and accounts for the cumulative effect of all characteristics that can affect likelihood. Otherwise, the features of the facility or process design that can affect the likelihood should be described. Regardless of the method used to assign a likelihood score, care must be taken that all facility and process features that can affect the event likelihood (reliability and availability qualities) are recognized as such and appropriately maintained. Appendix C to Chapter 3 contains additional guidance applicable to internal initiating events other than IROFS failures.

Similarly, the likelihood of internal initiating events that are IROFS failures also depends on specific, identifiable characteristics of the facility or process design. Scores may be assigned to such events based either on objective evidence of their frequency of occurrence or on specific identifiable characteristics of the IROFS that can affect the frequency of occurrence. If the actual frequency of occurrence is known, this information should be used. Otherwise, the features of the IROFS that can affect the likelihood should be described. Regardless of the method used to assign a likelihood score, care must be taken that all IROFS attributes that can affect the event likelihood (reliability and availability qualities) are recognized as such and appropriately maintained. The following provides guidance on specific reliability and availability qualities associated with individual IROFS.

For both types of internal initiating events, facility or process features (or physical and chemical phenomena) that can affect the initiating event likelihood may be identified as initial conditions or bounding assumptions. The important factor is that these initial conditions and bounding assumptions must be identified and, if susceptible to change over the lifetime of the facility (such as through process deviations or facility changes), must be appropriately maintained. For example, the maximum throughput or inventory in a process may change; thus, measures should be in place to maintain this throughput or inventory if it is relied on to meet the performance requirements, whereas the flow of gravity or maximum density may not require specific controls.

**Individual IROFS**

Section 3.4.3.2(9) of Chapter 3 of this NUREG states that the reliability and availability qualities of individual IROFS include (1) safety margin in the controlled parameter, (2) the type of IROFS (passive or active engineered, simple or enhanced administrative), (3) the type and safety grading of any management measures, (4) whether the system is fail-safe, failure is self-announcing, or the IROFS is subject to periodic surveillance, (5) failure modes, (6) demand rate, and (7) failure rate. It is very important that any qualitative (or quantitative) method of likelihood evaluation consider all applicable IROFS attributes that could affect the reliability and availability of the IROFS, such as those discussed below. For example, reliance should not be based solely on the type of IROFS (passive engineered, active engineered, simple administrative, or enhanced administrative).

In addition to those reliability and availability qualities discussed above, other factors may require consideration. For example, environmental conditions, such as extreme temperatures and pressures, corrosive atmosphere, excessive vibration, may have a significant effect on IROFS reliability and should be appropriately considered.

The level of detail describing the IROFS in the ISA Summary is also important. It would be acceptable to describe the IROFS at the system level if that is sufficient to demonstrate compliance with the performance requirements. The regulation in 10 CFR 70.65(b)(6) states that IROFS should be described "in sufficient detail to understand their functions in relation to the performance requirements." It is important that the description be sufficiently detailed to identify all attributes of the IROFS that can affect its likelihood of failure, as well as everything that is within the boundary of the IROFS. It may not be necessary to specify the model number or exact design of a pump if the only attribute relied on to meet the performance requirement is the pumping capacity or oil reservoir volume. It may be sufficient to describe the pump as "centrifugal pump limited to less than 10 liters oil." The IROFS boundary includes everything necessary for the IROFS to perform its intended safety function. For example, the boundary of an enhanced administrative IROFS includes all instrumentation (sensors, annunciators, circuitry, any controls activated by the operator) relied on to trigger the operator action; the boundary of a simple administrative control includes the equipment necessary to correctly perform the action; and the boundary of an active engineered control includes the attendant instrumentation, sensors, essential utilities, and any auxiliary equipment needed to perform its safety function. The reliability and availability qualities of every component within the IROFS boundary must be considered in evaluating the total IROFS likelihood.

Additional guidance on some of the specific reliability and availability qualities of individual IROFS is provided below.

Safety Margin in Controlled Parameter: "Safety margin" refers to the difference between the value of a parameter likely to be encountered during normal or credible abnormal conditions and the value that would allow an accident to be possible. The precise value of the margin in terms of the parameter is not meaningful; rather, for the event to be unlikely or highly unlikely based on safety margin, the margin should be several times larger than the expected process variation or uncertainty. Similarly, if the margin is much greater than the change in the parameter resulting from the worst case credible upset, this fact could be credited for ensuring that the event is unlikely or highly unlikely.

The phrase "controlled parameter" indicates that means should be provided to ensure that the safety margin is continuously present, if the margin is relied on in evaluating likelihood. Parameters that are not controlled should be considered to be at their worst case credible values.

Type of Control: Passive engineered controls are generally considered preferable to active engineered controls, active engineered controls preferable to enhanced administrative controls, and enhanced administrative controls preferable to simple administrative controls. This is because, ordinarily, passive engineered controls are the most reliable, and simple administrative controls are the least reliable. Although this is one of the factors that should be considered, evaluations of likelihood should not rely solely on the type of control. This is because the likelihood associated with passive engineered controls, for example, can vary widely depending on specific attributes of the IROFS.

Type and Safety Grading of Management Measures: The specific management measures applied to an IROFS can have a significant effect on its overall likelihood. Of particular importance is surveillance, because this can have a direct and transparent effect on the duration of failure in a method that gives credit to duration of failure. It may not be necessary to specify the frequency of preventive maintenance, testing, and calibration quantitatively in the ISA Summary. For example, to take credit for generic failure rates for a piece of equipment, it may be sufficient to specify that maintenance will be performed at a frequency and in a manner consistent with the manufacturer's recommendations. Functional testing should be conducted in a manner that ensures that everything within the IROFS boundary is working as needed for the IROFS to perform its safety function.

While the degree and type of management measures can increase or decrease the likelihood score associated with an IROFS, primary reliance should be on designing IROFS that have a certain reliability and then applying management measures to maintain that reliability. It should not be supposed that one can achieve any desired reliability by applying increasingly stringent management measures.

Fail-Safe or Self-Announcing: This is the characteristic of an IROFS that determines the degree to which failure of an IROFS is detected and appropriately corrected. For the purpose of the ISA and ISA Summary, an IROFS is considered to fail only when it fails to perform its intended safety function. Thus, a valve that is an IROFS is not considered to fail in the context of the accident sequence (i.e., to contribute to the progression of an accident sequence) as long as it fails in a safe configuration (fails-safe). If the valve is designed to fail closed (and closed is the safe configuration), credit may be taken for the fact that the valve is designed to fail closed. The likelihood thus is not the likelihood that the valve fails, but the likelihood that it fails in a way other than how it is designed to fail. An IROFS that is fail-safe may include within its boundary a system designed to put the process into a safe condition upon failure of a component. An IROFS whose failure is self-announcing is one in which failure is either self-revealing (e.g., by presence of solution on a floor where operators are continuously present) or results in an alarm to alert operators. The main effect for the ISA Summary is to limit the duration of failure by ensuring that the upset condition is corrected essentially immediately. Similarly, surveillance may be relied on to limit the duration of failure to a specified period.

Failure Modes: In addition to specifying the safety function that an IROFS must perform, it is necessary to consider the specific failure modes of the IROFS. A particular IROFS may be credited in several different accident sequences but may have different scores in each because of the differing failure modes leading to an accident. For example, a pipe may either plug or leak. A valve may leak, fail open, or fail closed. A complex piece of equipment such as a pump may have multiple different failure modes, each with a different likelihood, leading to several different accident sequences. The description of the accident sequence should clearly specify the conditions and failures that must occur for the undesired consequences to result.

Demand Rate: Demand rate refers to the frequency with which an IROFS having a specified probability of failure on demand is required to perform its safety function. The number of times an IROFS is required to work can have a significant effect on its likelihood of failure. For example, a particular administrative control may have a certain failure likelihood. However, whether the accident sequence is "not unlikely," "unlikely," or "highly unlikely" will depend on the frequency with which the action is performed. If the action is required several hundred times a year, then occurrence of the initiating event will be significantly more likely than if the action is required once per year. Similarly, a passive control (such as the integrity of a storage container) may have a certain failure likelihood. However, if there are a thousand such containers in a storage array, then the likelihood that any one container will leak is much greater than if there is only one such container. Care must be taken to specify whether the initiating event is the leak of a particular container, or any one container, in the array.

Failure Rate: Failure rate refers to the frequency with which a continuously demanded item is observed to fail. In a qualitative method for likelihood evaluation, the failure rate is described in terms of qualitative descriptors (e.g., "several failures per year," "a few failures during facility lifetime," "no failures in 30 years for tens of similar IROFS in industry") used in the assignment of qualitative likelihood scores (e.g., -1, -2, -3; A, B, C). This information is often not available with any precision, but when available, it should be used along with other qualitative information in the assignment of scores. This is because the failure rate represents an objective measure of the cumulative effect of all the reliability and availability qualities of the system. (See the discussion of qualitative and quantitative information below.)

This is not intended to be a comprehensive list of all facility- or process-specific factors that can affect the failure likelihood of individual IROFS.

## Accident Sequences

Section 3.4.3.2(9) of this SRP states that there are other reliability and availability qualities that relate to characteristics of the entire system of IROFS credited in the accident sequence. This is because the accident sequence likelihood is not just a function of the likelihood of failure of the individual IROFS, but also of the relationship between the IROFS.

Additional guidance on some of the specific reliability and availability qualities applicable to the accident sequence as a whole is provided below.

Defense in Depth: Defense in depth is the degree to which multiple IROFS or systems of IROFS must fail before the undesired consequences (e.g., criticality, chemical release) can result. IROFS that provide for defense in depth may be either independent or dependent, although IROFS should be independent whenever practical because of the possibility that the reliability of any single IROFS may not be as great as anticipated. This will make the results of the risk evaluation more tolerant of error. In addition, IROFS must be independent if the method for likelihood determination assumes independence (such as methods relying on summation of indices). IROFS are independent if there is no credible single event (common-mode failure) that can cause the safety function of each IROFS to fail. Multiple independent IROFS generally provide the highest level of risk reduction. The degrees of redundancy, independence, and diversity are important factors in determining the amount of risk reduction afforded by the system of IROFS.

Degree of Redundancy: Defense in depth is provided by specifying redundant IROFS that perform the same essential safety function. Redundant IROFS may be either diverse or nondiverse; it is not necessary for them to consist of identical equipment or operator actions. However, when identical equipment or operator actions provide redundancy, it is important to ensure that all credible common-mode failures have been identified.

Degree of Independence: To qualify as independent, the failure of one IROFS should neither cause the failure nor increase the likelihood of failure of another IROFS. No single credible event should be able to defeat the system of IROFS such that an accident is possible. A systematic method of hazard identification should thus be used to provide a high degree of assurance that all credible failure mechanisms that could contribute to (i.e., by initiating or failing to prevent or mitigate) an accident have been identified. Methods commonly used for likelihood evaluation almost always assume that the chosen IROFS are independent. Examples of these methods include layer of protection analysis and the index method in Appendix A to this report. In a few cases, it may not be feasible to entirely eliminate the possibility of dependent failures. Methods that rely on independent IROFS should not be used to evaluate the likelihood of systems of IROFS with dependent failures. (Guidance applicable to the rare system with dependent failures is provided below.) If, however, the common-cause failure is sufficiently unlikely, it may be possible to treat IROFS as independent for purposes of the ISA and ISA

Summary, as discussed below. Because of the added requirement to meet the double-contingency principle, this approach will not be valid for criticality accident sequences when the requirements of 10 CFR 70.64(a)(9) apply.

Many factors can lead to IROFS not being independent, and these factors can have a significant effect on the likelihood of an accident sequence. A partial list of conditions that will almost always lead to two or more IROFS not being independent follows:

- The same individual performs administrative actions.

- Two different individuals perform administrative actions but use the same equipment and/or procedures.

- Two engineered controls share a common hardware component or common software.

- Two engineered controls measure the same physical variable using the same model or type of hardware.

- Two engineered controls rely on the same source of essential utilities (e.g., electricity, instrument air, compressed nitrogen, water).

- Two engineered controls are collocated such that credible internal or external events (e.g., structural failure, forklift impacts, fires, explosions, chemical releases) can cause both to fail.

- Administrative or engineered controls are susceptible to failure because of the presence of credible environmental conditions (e.g., two operator actions defeated by corrosive atmosphere, sensors rendered inoperable because of high temperature).

The presence of any of these conditions does not necessarily mean that the IROFS cannot be considered independent, but the applicant should provide additional justification demonstrating the lack of common-mode failure. The likelihood of such conditions in relation to the overall likelihood of an accident should be factored into the determination of the significance of the common-mode failure.

Diversity: Diversity is the degree to which IROFS that perform different safety functions provide defense in depth. This means that different types of failures must occur before an accident is possible. Diverse controls may consist of controls on different parameters or different means of controlling the same parameter. In choosing redundant controls, preference should be given to diverse means of control, because they are generally less susceptible to common-mode failure than are nondiverse means. However, it is still necessary to consider all credible failure modes of the system when evaluating the overall likelihood of failure.

Vulnerability to Common-Cause Failure: Diverse means of control should be provided whenever practicable to minimize the potential for common-mode failure. For example, Chapter 5 of this SRP states that for criticality protection, a two-parameter control should be considered preferable to two controls on one parameter. Where a two-parameter control is not practicable, diverse means of controlling a single parameter should likewise be considered preferable to two redundant controls on that single parameter.

It is not always possible to provide absolute assurance that IROFS are perfectly independent. However, if the cumulative likelihood of all common-mode failures of a system of IROFS is significantly less than the independent failure of the system of IROFS, then the IROFS may be treated for all practical purposes as independent. Quantitatively, this means that the likelihood of the common-cause failure should be at least two orders of magnitude less than that of the independent failure of the system of IROFS. Qualitatively, this means that the likelihood of the common-cause failure should be sufficiently low that it does not change the score for the system of IROFS.

If credible common-mode failures cannot be neglected, as discussed above, then they must be considered in evaluating the overall accident sequence likelihood. A likelihood evaluation method (whether quantitative or qualitative) that correctly treats dependent failures should be used when such failures are present.

In general, the probability of failure of a system of two IROFS may be expressed as:

$$P(A,B) = P_{ind}(A,B) + P_{dep}(A,B) = P(A)P(B) + P_{dep}(A,B)$$

That is, there is a component to the likelihood that is the independent failure of IROFS A and B and a component that represents the common-mode failure of IROFS A and B. Independent failure of the IROFS is represented by the product P(A)P(B). Therefore, the condition that the two IROFS be considered independent may be expressed as:

$$P(A,B) \approx P(A)P(B)$$

or equivalently

$$P_{dep}(A,B) << P(A)P(B)$$

A variety of different methods may be used to treat dependent failures when the conditions above are not met. For example, in a quantitative method, the likelihood of the common-mode event may be estimated and factored into the above equation. In a qualitative scoring method, the likelihood score may be increased to reflect the existence of a common-mode failure. (In a qualitative scoring method similar to that employed in Appendix A to Chapter 3 of this SRP, summation of individual IROFS scores to determine the overall accident sequence score is permissible only if the IROFS are independent. Such a method assumes that independence should be modified as needed to correctly treat common-mode failures.) In the layer of protection analysis method, only the independent IROFS are credited in evaluating the overall accident sequence likelihood. In a qualitative fault tree method, the common-mode failure may be included as an additional basic event in the fault tree. It is permissible then to treat the independent failure of the system of IROFS as one accident sequence and the dependent failure as another. The method used to treat dependent failures should be appropriately justified.

Qualitative criteria may be used to assess the effect of dependent failures on likelihood scores. The effect of qualitative performance-shaping factors should be considered in these criteria. For example, repeated failures of identical administrative IROFS (e.g., multiple batching, multiple valving, or spacing violations) should not be considered to be independent nor receive the same score without substantial justification, as discussed below. This is because the likelihood of subsequent human failures increases once the initial failure has occurred. The set of factors that could contribute to multiple administrative failures may include inadequate or out-of-date procedures, poor training, environmental distractions, and poor human factors design. For the same reason, the possibility of two different administrative failures by the same individual should be carefully considered for common-mode vulnerability. In assessing the vulnerability of these actions to common-mode failure, consideration may be given to any recovery factors that may be in place to interrupt the sequence of failures (e.g., supervisory checking, inspection, independent verification). Such recovery factors should be treated as measures that enhance the reliability of the administrative IROFS or ensure that repeated failures may be considered to be independent. In particular, independent verification of one administrative IROFS should not be used as a separate IROFS in the same accident sequence. For the same reasons as cited above, verification that an action has been performed correctly would be susceptible to the same factors that caused the initial failure. In addition, verification of an action is likely to be more cursory and, therefore, less reliable than performance of the original action. Moreover, in the event that the first action was performed correctly, the independent verification of that first action would not contribute to meeting the performance requirements, and therefore, the first action would constitute a sole IROFS. Thus, independent verification should be used only to increase the reliability of an IROFS and should not be treated as a separate IROFS nor credited with the same level of risk reduction.

In addition to the above, for criticality accident sequences required to comply with the double-contingency principle (see appendix 5-A of this SRP).

**Use of Quantitative and Qualitative Information**

Section 3.4.3.2(9) of this SRP acknowledges that a mix of quantitative and qualitative information is often available to an analyst performing an ISA. This SRP includes a list of some types of objective quantitative information and states that this information should be considered in evaluating likelihood, even in purely qualitative methods. The information listed includes (1) reports of equipment failures or procedural violations, (2) surveillance intervals, (3) functional testing intervals or audit frequencies, (4) time required to render the system safe, and (5) demand rates. In a purely qualitative method, such information, to the extent it is available, should be considered qualitatively. One example of this is using surveillance periods as part of the justification for qualitative duration indices (as in Appendix A to Chapter 3 of this SRP).

In using such objective data, facility-specific data are preferable to generic data, and process-specific data are preferable to facility-specific data because of the many environmental and other factors that can affect likelihood. For example, a manufacturer may have certified a particular pump with a given reliability rating, but the actual performance in process will depend on maintenance, electrical and mechanical loading, type of oil, ambient temperature, and vibration, among other factors. While more specific data are preferable, typically, the more specific the conditions, the fewer data are available. The amount and specificity of the data should be given appropriate weight in evaluating likelihood. For example, the use of generic failure data for a specific type of valve may be acceptable if an appropriately bounding value (i.e., the less conservative extreme of a range of values) is used. A less bounding value may be acceptable if information is available from the manufacturer on the specific model of valve. An even less bounding value may be acceptable if sufficient operating experience is available to support facility- or process-specific values. Sufficient margin to bound uncertainties in failure rates should be provided when relying on generic information.

Operating history may be credited in justifying likelihood scores for individual IROFS. Care must be taken that this credit is based on documented performance data and not anecdotal evidence and that the operating history is applicable to the event being evaluated. For example, not having any criticality accidents in 30 years of operation would not be justification for a failure frequency for a particular component or initiating event (since the initiating event may have occurred several times during that time period without resulting in a criticality). It would also not be justification for a likelihood corresponding to a time between failures longer than 30 years. In addition, if significant facility changes occurred over the previous 30 years of operation, this information may not be meaningful. The limits and applicability of the operating data used to justify likelihood should be explained.

Especially for new processes or facilities, such objective quantitative data may not be available. Appropriate margin in plant operations and conservatism in likelihood scoring should be used and justified when such information is not available. Over the facility lifetime, however, information gained with regard to operational events and IROFS failures should be evaluated and fed back into the ISA process. This may be justification for reducing margins and conservatism over the facility lifetime.

**Graded Approach to Integrated Safety Analysis**

The performance requirements of 10 CFR 70.61(b) and (c) establish an acceptable level of risk, in that high-consequence events must be made "highly unlikely" and intermediate-consequence events must be made "unlikely." In addition, 10 CFR 70.65(b)(4) requires that an applicant's ISA Summary contain a demonstration of compliance with the performance requirements of 10 CFR 70.61. The means and the level of effort required to demonstrate compliance with 10 CFR 70.61 depend on the amount of risk reduction needed to meet the likelihood thresholds in that regulation. For example, a facility that obviously has inherently low risk (even before the performance of the ISA) requires less effort to demonstrate compliance than an inherently higher risk facility. Examples would include facilities with small mass or very low enrichment of special nuclear material, low chemical inventories, or insignificant combustible loading. Thus, the ISA methods used may be graded commensurate with the risk of the facility.

The facility and process characteristics that determine inherent risk should be identified as initial conditions and/or assumptions and appropriately identified and maintained to ensure that they will be present over the lifetime of the facility, if credit is taken for them in meeting the performance requirements. For example, a possession limit on the maximum enrichment or amount of special nuclear material at the facility may be credited in ensuring low risk of criticality, because the license sets an explicit limit. Chemical inventories may be likewise credited, provided that they are limited by license or the maximum inventory is identified as important to safety and rigorously controlled. ISA methods may be graded commensurate with the amount of risk reduction required once these factors have been explicitly identified and maintained.

The following are examples of aspects of the ISA process that may be graded commensurate with risk:

- In the selection of the hazard identification method, the what-if or what-if/checklist method would be more suitable for low-risk, simple operations; hazardous operations, fault tree, and other sophisticated methods may be appropriate for more complex or higher risk operations.

- In the evaluation of the type, number, and robustness of IROFS, lower risk facilities will not require the same level of control.

- In the application of management measures, lower risk facilities will not require measures as stringent as those for higher risk facilities.

- In the evaluation of likelihood, the technical justification required to support a high degree of risk reduction is much greater than that required to support a low or moderate degree of risk reduction. Methods used to support a high degree of risk reduction should be more sophisticated, and warrant greater regulatory scrutiny, than methods used to support a lower degree of risk reduction.

In addition to the inherent risk of the facility or process, the amount of conservatism may be considered in grading ISA methods. For example, if a very conservative likelihood is assumed for all IROFS failures, then the rigor and level of detail in describing the IROFS, considering all reliability and availability qualities and treating dependent failures, would not have to be at the same level as in a facility taking more realistic credit for IROFS failures. The grading of ISA methods necessitates that the applicant demonstrate (1) that the risk is inherently low and will be maintained over the lifetime of the facility, or (2) that there is a consistent and dependable amount of conservatism in ISA methods that offsets the uncertainty arising from lack of rigor.

## Regulatory Basis

The risk of each credible high-consequence event must be limited. Engineered controls, administrative controls, or both shall be applied to the extent needed to reduce the likelihood of occurrence of the event so that, upon implementation of such controls, the event is highly unlikely or its consequences are less severe than those described in 10 CFR 70.61(b)(1–4).

The risk of each credible intermediate-consequence event must be limited. Engineered controls, administrative controls, or both shall be applied to the extent needed so that upon implementation of such controls, the event is unlikely or its consequences are less than those described in 10 CFR 70.61(c)(1–4).

Each licensee or applicant shall conduct and maintain an ISA that is of appropriate detail for the complexity of the process and that identifies "the consequences and likelihood of occurrence of each potential accident sequence…and the methods used to determine the consequences and likelihoods," as stated in 10 CFR 70.62(c)(1)(v).

The ISA Summary must contain "information that demonstrates the licensee's compliance with the performance requirements of Section 70.61," as stated in 10 CFR 70.65(b)(4).

The ISA Summary must also include the definitions of "unlikely," "highly unlikely," and "credible" as used in the evaluations of the ISA, as stated in 10 CFR 70.65(b)(9).

## Technical Review Guidance

The reviewer should use the information contained in this appendix, as applicable, to evaluate an applicant's or a licensee's qualitative methods of likelihood evaluation, commensurate with the level of risk reduction required to comply with the performance requirements of 10 CFR 70.61. If the applicant is using the index method defined in Appendix A to Chapter 3 of this SRP, the reviewer should use the guidance in Appendix A to evaluate the adequacy of the applicant's ISA Summary. The purpose of the ISA Summary review is not to verify the correctness of the likelihood scores for every single accident sequence, but to verify that the applicant has an acceptable methodology that contributes to reasonable assurance of maintaining an adequate safety basis over the facility lifetime, by ensuring that the methodology results in assignment of appropriate likelihoods. Thus, the reviewer should primarily determine whether there is a justifiable basis for the scores, and whether there is reasonable assurance that this basis will be maintained over the facility lifetime, assuming the application of appropriate management measures.

The applicant's qualitative method for likelihood evaluation should be acceptable if the following are true:

- The definitions of likelihood are clear, are based on objective criteria, and can consistently distinguish events in different likelihood categories.

- The methods for likelihood evaluation are consistent with the likelihood definitions and the process being evaluated (e.g., the methods correctly treat initiating events and initial conditions, subsequent failures, and dependent failures).

- The methods for likelihood evaluation appropriately consider all availability and reliability qualities of individual IROFS and the interdependencies between them in assigning qualitative likelihood scores.

- The ISA Summary describes initiating events, initial conditions, and subsequent IROFS failures in detail sufficient to demonstrate that the performance requirements will be met and maintained.

## Recommendations

This guidance should be used to supplement Chapter 3 and Appendix A to this SRP.

## References

*U.S. Code of Federal Regulations*, Title 10, "Energy," Part 70, "Domestic Licensing of Special Nuclear Material."

U.S. Nuclear Regulatory Commission, "Standard Review Plan for the Review of an Application for a Mixed Oxide (MOX) Fuel Fabrication Facility," NUREG-1718, August 2000.

# APPENDIX C

# INITIATING EVENT FREQUENCY

## Purpose

This appendix addresses the measures needed to ensure the validity and maintenance of the initiating event frequencies (IEFs) used to demonstrate compliance with Title 10 of the *Code of Federal Regulations* (10 CFR) 70.61, "Performance Requirements."

## Introduction

The purpose of this appendix is to clarify the use of IEFs for demonstrating compliance with the performance requirements of 10 CFR 70.61. NUREG-1718, "Standard Review Plan for the Review of an Application for a Mixed Oxide (MOX) Fuel Fabrication Facility," issued August 2000, and this Standard Review Plan (SRP) provide methods for reviewing integrated safety analyses (ISAs) by employing a semi-quantitative risk index method. While one of these methods is described below to illustrate the use of IEFs, applicants and licensees may use other methods that would produce similar results. No particular method is explicitly mandated, and sequences that are risk significant or marginally acceptable are candidates for more detailed evaluation by the applicant or licensee and reviewer.

## Discussion

Each licensee or applicant is required to perform an ISA to identify all credible high-consequence and intermediate-consequence events. The risk of each such credible event is to be limited through the use of appropriate engineered and/or administrative controls to meet the performance requirements of 10 CFR 70.61. Such a control is referred to as an item relied on for safety (IROFS). In turn, a safety program must be established and maintained to ensure that each IROFS is available and reliable to perform Its intended function when needed. The safety program may be graded such that the management measures applied are graded commensurate with the reduction of risk attributable to that item. In addition, a configuration management system must be established pursuant to 10 CFR 70.72, "Facility Changes and Change Process," to evaluate changes and to ensure, in part, that the IROFS are not removed without at least equivalent replacement of the safety function.

The risk of each credible event is determined by cross-referencing the severity of the consequence of the unmitigated accident sequence with the likelihood of occurrence in a risk matrix with risk index values. The likelihood of occurrence risk index values can be determined by considering the criteria in Tables A-9 through A-11 in Appendix A to Chapter 3 of this SRP. Accident sequences result from initiating events that are followed by the failure of one or more IROFS. Initiating events can be (1) an external event such as a hurricane or earthquake, (2) a facility event external to the process being analyzed (e.g., fires, explosions, failures of other equipment, flooding from facility water sources), (3) deviations from normal operations of the process (credible abnormal events), or (4) failures of an IROFS in the process. (Appendix D to Chapter 3 offers additional guidance regarding initiating probabilities from natural phenomena hazards.)

An initiating event does not have to be an IROFS failure. An item becomes an IROFS only if the ISA credits it for mitigation or prevention per the definition in 10 CFR 70.4, "Definitions." If an item whose failure initiates an event has strictly an operational function, it does not have to be an IROFS. This applies to external events and can apply to internal events. If the item whose failure initiates an event has solely a safety function that is credited in the ISA, then it should be an IROFS. If the item has both an operational and a safety function, the safety function should make it an IROFS (for its ISA-credited safety features only).

IEFs can play a significant role in determining whether the performance requirements of 10 CFR 70.61 are met for a particular accident sequence. Whether an initiating event results from an IROFS or a non-IROFS failure, licensees should take appropriate action to ensure that any change to the basis for assigning an IEF value to that event is evaluated on a continuing basis to ensure continued compliance with the performance requirements. For example, a non-IROFS component may not be subject to the same quality assurance (QA) program controls and other management measures that an IROFS would receive (i.e., surveillance, testing, procurement). However, appropriate management controls should be considered, in a graded manner, to provide assurance that performance requirements are met over time. The ability to identify a non-IROFS component failure, similar to that for IROFS, may be needed to provide feedback on failure rates and IEFs to the ISA process. Changes to the IEF values may result from changes to a component's design, procurement, operation, or maintenance history, as well as new or increased external plant hazards, and should be considered in a graded approach.

## Regulatory Basis

This guidance relies on the following regulatory bases:

- 10 CFR 70.61
- 10 CFR 70.62, "Safety Program and Integrated Safety Analysis"
- 10 CFR 70.65, "Additional Content of Applications"
- 10 CFR 70.72, "Facility Changes and Change Process"

## Applicability

This guidance is for use in those cases where an applicant or licensee chooses to use an IROFS or non-IROFS failure IEF for risk determination.

## Technical Review Guidance

1. Initiating Event Frequency and Identification of an IROFS

Example

A licensee uses a heater/blower unit to heat a uranium hexafluoride ($UF_6$) cylinder in a hot box to liquefy the contents before sampling. The unmitigated accident sequence involves the failure of the controller for the heater/blower resulting in overheating of the cylinder. This results in the cylinder becoming overpressurized and rupturing, which releases the $UF_6$ to the surrounding process area. Analysis of such a release indicates that it would exceed the performance requirements of 10 CFR 70.61. The licensee has two basic choices: (1) assume that the initiating event probability equals 1 and provide an appropriate level of mitigation or prevention solely through one or more IROFS or (2) assign a value to the initiating event (blower/heater controller failure) and provide one or more preventive or mitigative IROFS to bring the accident sequence risk within the performance requirements.

If the licensee chooses the second option and assigns an appropriate value to the IEF, the indices of Table A-9 in Appendix A to Chapter 3 of this SRP may be used. The controller for the heater/blower unit would be assigned an appropriate frequency index number. The licensee would then analyze the accident sequence and determine whether additional IROFS are necessary to meet the performance requirements. There are now two variables that feed into the risk determination: one or more IROFS controllers for the heater/blower unit in a manner that changes the licensee's previous determination of compliance with the performance requirements must be evaluated per 10 CFR 70.72(a).

2.    Initiating Event Frequency Index Use

Indices may be used to determine the overall likelihood of an accident sequence. Table A-9 of Appendix A to Chapter 3 of this SRP identifies frequency index numbers based on specified evidence. The evidence used by applicants and licensees should be supportable and documented in the ISA Summary as required by 10 CFR 70.65(b)(4). The evidence cited in the ISA documentation should not be limited to anecdotal accounts and must demonstrate compliance with the definitions of "unlikely," "highly unlikely," and "credible" as required by 10 CFR 70.65(b)(9). The rigor and specificity of the documented evidence should be commensurate with the item's importance to safety, and the data should support the frequency chosen (e.g., data from 30 years of plant operating experience based on a single component typically could not be expected to support a $10^{-2}$ failure probability).

An item's failure rate should be determined from actual data for that specific component or safety function in the current system design under the current environmental conditions. When specific failure data are limited or not available, the applicant or licensee may use more "generic" data with appropriate substantiation. However, when less specific failure data are available, appropriate conservatism should be exercised in assigning frequency indices. The footnote to Table A-9 that states "Indices less than (more negative than) -1 should not be assigned to IROFS unless the configuration management, auditing, and other management measures are of high quality, because without these measures, the IROFS may be changed or not maintained" should also be applied to non-IROFS IEFs. In this case, appropriate management controls should be provided to ensure that any changes to the evidence supporting IEF indices will be identified and promptly evaluated to ensure that the performance requirements of 10 CFR 70.61 are met. A graded approach may be used in applying management controls based on the IEF values; however, the ISA Summary should explain how this will be done.

The licensee or applicant should periodically evaluate possible changes to IEFs, failure rates, and the assumptions they are based on to ensure that the ISA process has accounted for any change to an IEF. Over time, an IEF may change because of component aging or deterioration. Maintenance and performance experience should be fed back into the IEF evaluation. IEF changes could involve, for example, the introduction of new effects or hazards from nearby processes or new materials or changes in design, maintenance, or operation activities. The applicant or licensee should establish management measures, which may be graded, to periodically confirm that the ISA assumptions have not changed. For example, an applicant or licensee may choose to verify that there have been no changes to hazards from maintenance activities during a certain period of time based on an appropriate documented technical review or audit under the QA program.

Whatever strategy the applicant or licensee chooses should result in timely identification and periodic evaluation of failure rates, followed by a prompt evaluation of the failure rate change on the ISA assumptions. This can be accomplished in accordance with the corrective maintenance program and/or the QA problem identification and corrective action system.

Indices particularly relied on (i.e., less than -1) for overall likelihood will be examined during the ISA review process.

3.    External Initiating Event Frequencies

The applicant or licensee should periodically evaluate possible changes to nonnatural phenomena external events to ensure that the ISA process has accounted for any change to an IEF. Such changes could involve, for example, the introduction of new hazards from an adjoining industrial site or changes in adjoining transportation activities. The applicant or licensee should establish management measures, which may be graded, to periodically confirm that the ISA assumptions have not changed. For example, an applicant or licensee may choose to verify that external hazards have not changed based on a 2- to 3-year review under the QA program.

4.    Assurance

The safety program required by 10 CFR 70.62(a) should have provisions for implementing the appropriate management controls to maintain the validity of the IEFs. Consideration should also be given to commitments in the QA program or a specific license condition.

## References

*U.S. Code of Federal Regulations*, Chapter I, Title 10, "Energy," Part 70, "Domestic Licensing of Special Nuclear Material."

U.S. Nuclear Regulatory Commission, "Standard Review Plan for the Review of an Application for a Mixed Oxide (MOX) Fuel Fabrication Facility," NUREG-1718, August 2000.

# APPENDIX D

# NATURAL PHENOMENA HAZARDS

## Purpose

This appendix provides additional guidance addressing accident sequences that may result from natural phenomena hazards in the context of a license application or an amendment request under Title 10 of the *Code of Federal Regulations* (10 CFR) Part 70, "Domestic Licensing of Special Nuclear Material," Subpart H, "Additional Requirements for Certain Licensees Authorized To Possess a Critical Mass of Special Nuclear Material."

## Introduction

This appendix provides additional guidance for reviewing the applicant's (or licensee's) evaluation of natural phenomena hazards up to and including "highly unlikely" events for both new and existing facilities.

## Discussion

For facilities processing special nuclear materials, 10 CFR 70.61, "Performance Requirements," requires that individual accident sequences resulting in high consequences to workers and the public be "highly unlikely" and that sequences resulting in intermediate consequences to these receptors be "unlikely." Although the regulations establish the threshold levels that differentiate high-consequence events from intermediate-consequence events, they do not define "highly unlikely" and "unlikely." According to 10 CFR 70.65(b)(9) and subject to staff approval, the integrated safety analysis (ISA) summary submitted by applicants and licensees must include definitions of these terms. Chapter 3 of this NUREG further describes the acceptance criteria for the definitions of these terms.

The implementation of these requirements may vary somewhat because of different definitions of likelihood proposed by different applicants (or licensees).[1] The regulation specifies quantitative consequence thresholds of the performance requirements (except for chemical releases). The regulation and its performance requirements pertain to existing facilities, as well as proposed facilities, and apply to manmade external hazards and natural phenomena hazards, in addition to process hazards. However, new facilities and new processes at existing facilities must also address the requirements of 10 CFR 70.64, "Requirements for New Facilities or New Processes at Existing Facilities," which includes the baseline design criterion for natural phenomena hazards (10 CFR 70.64(a)(2)). This baseline design criterion requires that "the design must provide for adequate protection against natural phenomena with consideration of the most severe documented historical events for the site." The Statement of Considerations describes the application of the baseline design criteria as consistent with good engineering practice, which dictates that certain minimum requirements should be applied to design and safety considerations. The baseline design criteria must be applied to the design of new facilities and new processes at existing facilities but does not require retrofits to existing facilities

---

[1] For natural phenomena, deterministically defined events such as the probable maximum flood (PMF) or safe-shutdown earthquake (SSE), which are used as reactor design bases, can also be applied to 10 CFR Part 70 facilities as "highly unlikely" events. The actual probability (or likelihood) of such events may be difficult to define quantitatively and varies from site to site.

or existing processes (e.g., those housing or adjacent to the new processes). Also included in 10 CFR 70.64(b) are a requirement for incorporation of defense in depth in design and a requirement to prefer engineered controls over administrative controls.

New structures associated with facilities being reviewed, such as the gas centrifuge facilities and the mixed oxide fuel fabrication facility, will be designed and constructed to meet the seismic regulatory requirements. Hence, these facilities and additional new facilities to be licensed under 10 CFR Part 70 are not expected to present designs with seismic deficiencies. New facilities can also be expected to be sited above a "highly unlikely" flood such as the PMF and can be expected to withstand tornado winds and missiles, if necessary.

Most structures at existing nuclear fuel cycle facilities are built to a model building code, which includes meeting a design-basis earthquake having an exceedance probability of $2 \times 10^{-3}$ per year to less than $10^{-3}$ per year (U.S. Department of Energy (DOE) Standard-1020-2002, "Natural Phenomena Hazards Design and Evaluation Criteria for Department of Energy Facilities," Appendix C). Existing facilities are generally sited above the 100-year flood plain and are designed for wind as well as snow and ice loading as specified in applicable building codes. Extreme natural events such as "highly unlikely" floods and/or earthquakes have not been calculated for many existing sites, and it would be expensive and time consuming to do so.

The staff believes that many existing facilities can be shown to be in compliance with, or at least near compliance with, the performance requirements of the regulation by accounting for conservatisms in the seismic, flooding, and wind design of the facility. In addition, relatively minor engineered improvements and administrative measures may further enhance safety, at least with respect to the public and other offsite receptors.

**Seismic Hazards**

Potential damage to and/or failure of items relied on for safety (IROFS) as the result of ground movement and/or the seismic response of adjacent or interior IROFS must be considered in the ISA and ISA Summary accident sequence evaluations. Damage or failures that also should be considered include the following:

- seismic-induced failure of a facility component that is not an IROFS but that can fall and damage an IROFS (for example, a heavy load drop from a crane onto a container)

- displacement of adjacent IROFS during a seismic event causing them to pound together

- displacement of adjacent components resulting in failure of connecting pipes or cables which may cause flooding, fires, and/or releases of radiological or chemical materials

Seismic event evaluations should also consider potential multiple failure of IROFS (for example, multiple failures of tanks).

DOE has also recognized the difference between earthquake design probability and the probability that a safety component cannot perform its function. To quantify this difference, DOE has developed a risk reduction factor, R, as the ratio between the seismic hazard exceedance probability and the performance goal probability. Conservatism in nuclear facility design arising from factors such as use of prescribed analysis methods, specification of material strengths, and limits on inelastic behavior explains at least part of this apparent reduction in actual risk. Appendix C to DOE Standard-1020-2002 discusses this risk reduction factor.

For a consequence to affect the public or external site workers, licensed material or hazardous chemicals that could affect the safety of licensed material must be released through at least one, and often two, confinement barriers, such as the following:

- storage containers, glove boxes, tanks, or handling devices
- ventilation system dynamic confinement and filtration
- building structural shell

Criticalities, on the other hand, may result from the introduction of a moderator or loss of safe geometric control of confined materials.

By using risk reduction factors calculated for a facility and its specific components and/or estimating the degree of failure by comparison with the observed behavior of similarly constructed buildings during severe earthquakes, analysts can postulate reasonable scenarios. These scenarios may not release all the material at risk or present an unimpeded leak path to receptors. For example, some facilities might be able to show that, even in the case of an earthquake that is "highly unlikely," only certain types of containers or confinement systems are likely to be breached. If the amount of material contained in such containers is variable, then that probabilistic component may be factored into the overall likelihood of the accident sequence. If employing some of these mitigating considerations in the analysis requires reliance on special containers or procedures, then additional IROFS may also be needed. Another factor to consider is the likely rate of release based on the damage sustained. For example, some facilities may lose dynamic confinement but maintain building integrity. In some processes, radiologically and/or chemically hazardous material is held inside its primary containment at subatmospheric pressure. In these cases, even though the primary containments are inside a structure designed to withstand less than a "highly unlikely" earthquake, the subatmospheric conditions may be sufficient to limit both facility worker and offsite doses in the event of a greater earthquake. For example, an earthquake that results in limited subatmospheric containment losses may allow adequately trained workers to evacuate and/or take mitigative actions. The buildings containing cylinders of liquid uranium hexafluoride ($UF_6$) at gas centrifuge facilities are designed for a "highly unlikely" earthquake. In addition, some buildings at one of the proposed facilities are equipped with a seismically activated interlock (an IROFS) that will shut off the buildings' heating, ventilation, and air conditioning system during an event, thus limiting any leakage of $UF_6$ to the outside.

**Flooding Hazards**

Most fuel cycle licensees do not require large quantities of cooling water and, therefore, do not need to be located near large bodies of water. A site licensed under 10 CFR Part 70 does not need to meet prescriptive flood protection requirements but does have to meet the performance requirements for all credible events including flooding. A site meeting the flood protection requirements of a commercial reactor should be considered as being designed or located adequately to withstand a "highly unlikely" flooding event. Section 2.4 of NUREG-1407, "Procedural and Submittal Guidance for the Individual Plant Examination of External Events (IPEEE) for Severe Accident Vulnerability," issued June 1991, states that the design-basis flood (which for river sites is the PMF) as described in Regulatory Guide 1.59, "Design Basis Flooding for Nuclear Power Plants," is estimated to have an exceedance frequency of less than $10^{-5}$ per year. Sites that do not meet this level of protection can still meet the 10 CFR 70.61 performance requirements but must be considered on an individual basis.

In an evaluation of the effects of flooding on existing facilities, the following flood-related hazards should be considered:

- river flooding

    – inundation and hydrostatic loading
    – dynamic forces
    – wave action
    – sedimentation and erosion
    – ice loading

- upstream dam failures

    – inundation and hydrostatic loading
    – dynamic forces
    – erosion and sedimentation

- precipitation/local storm runoff

    – inundation (local ponding) and hydrostatic loading
    – dynamic loads (flash flooding)

- tsunami, seiche, hurricane storm surge

    – inundation and hydrostatic loading
    – dynamic forces
    – wave action

American National Standards Institute/American Nuclear Society Standard 2.8, "Determining Design Basis Flooding at Power Reactor Sites," issued July 1992, describes methods for determining these flooding and water-related effects for reactor sites. These methods can be applied to 10 CFR 70.61 analyses with less conservatism in some of these parameters.

A standard siting requirement for residential and commercial developments is to be above the 100-year flood plain. For large river basins, warning time and time to secure materials and evacuate personnel will probably be available. For small streams, there may be relatively little warning in regard to thunderstorms and localized rainfall. In such cases, rapid actions may be the only administrative protection available. An evaluation of the effectiveness of proposed protection will need to consider the effects of inundation, hydrostatic loading, erosion, and sedimentation. At a minimum, this would require that criticality events be prevented and materials remain confined within site structures.

At some sites, a delineation of the 500-year flood plain may also be available. If the site is above the 500-year flood plain, flooding may be considered an unlikely[2] event, depending on

---

[2]     Even if the licensee defines "unlikely" as less than $10^{-3}$ per year for the process sequences in the ISA summary, the conservative assumptions inherent in most flood plain hydrologic studies, such as those

the quality of the estimate. In this category, criticality events should still be prevented, but the breaching of a limited number of material containers may be allowable under the performance requirements (up to 25 rem for the public, up to 100 rem for workers, and a specified release limit) for events that, in terms of likelihood, are between "unlikely" and "highly unlikely."

In addition to the facility's location relative to the 100-year or 500-year flood plains, the effects of local intense precipitation and snow load should be considered. Local intense precipitation, especially in the form of snow, can result in roof collapse and localized site flooding. Normally, protection from local precipitation and snow is relatively easy to achieve through roof design and local site drainage design.

## Wind and Tornado Loading

Wind design for an existing facility if prescribed by an applicable building code would have an annual exceedance probability of greater than or equal to $2\times10^{-2}$. At such relatively high probabilities, tornado design criteria are not specified. However, depending on the geographic location of the facility, the effects of a tornado with an annual exceedance probability of $10^{-5}$ or greater may need to be considered.

Wind forces on walls of structures should be determined using appropriate pressure coefficients, gust factors, and other site-specific adjustments. If the wind is likely to blow inside the structure, either through design or wind-driven missile vulnerability, the effects of wind on internal IROFS requires consideration. If the winds are from a tornado, the effects of the atmospheric pressure change associated with the tornado must be considered. Normally, ventilation systems are most vulnerable to atmospheric pressure change, but windows, buried tanks, and sand filters can also be affected.

For straight winds, hurricanes, and weak tornadoes, missile criteria as specified in Table 3-3 of DOE Standard-1020-2002 may be considered. The missile specified is a 15-pound plank, measuring 2 inches by 4 inches, at a specified elevation and impact velocity. For facilities that may be subjected to more severe tornado missiles, the guidance in Tables 3-4 and 3-5 of DOE Standard-1020-2002 may be followed. For the tornado, a 3,000-pound automobile rolling and tumbling on the ground should also be considered. For such evaluations, the probability of the entire sequence should be considered, and missile criteria from either Table 3-4 or 3-5 of DOE Standard-1020-2002 may be used as appropriate.

## Considerations for Existing Processes at Existing Facilities

For existing processes at existing facilities, licensees are not required to address 10 CFR 70.64 baseline design criteria. However, they must still meet the performance requirements of 10 CFR 70.61, including accidents caused by natural phenomena, for which the staff may require additional IROFS to meet the performance requirements. Existing facilities can use IROFS in the form of additional administrative controls to meet the performance requirements without the need for design features normally required by accepted engineering practice. When near compliance can be achieved and complete compliance will be relatively costly, plants may request an exemption to the regulation.

---

performed for Federal Emergency Management Agency flood insurance rate maps, should justify the consideration of flooding above the 500-year flood plain as an unlikely event.

As discussed earlier, many existing 10 CFR Part 70 facilities are not designed for an earthquake beyond that specified in applicable building codes. Although this design may provide fairly good seismic protection to the structure, it may not protect internal equipment. Also, an existing facility may not be designed to any specific seismic criteria, in which case its ability to withstand earthquakes can only be estimated based on comparison with similar structures or through complex structural analysis. In such cases, licensees may add IROFS to meet the performance requirements. An example of such IROFS (procedures and upgrades) being effectively implemented would be a facility where the consequences of a release of licensed material to the public in a seismic event would be from fires and/or explosions. In this case, fixes such as seismically qualified flammable gas shutoff valves or electrical shutoffs might provide a large decrease in potential seismic consequences.

In regard to flooding, flood elevations beyond that of the 100-year flood may not have been determined for the site. For sites in proximity to a river, these determinations could be expensive and time consuming. For these cases, flood warning time may allow measures such as moving material at risk and/or blocking doors and openings in the facility structure.

A facility's ability to withstand high winds, rain and snow loads, and exterior fires can likewise be improved through a combination of administrative procedures and engineered improvements. Removing material at risk from under walls or roofs that are not seismically designed can reduce potential releases in case of collapse from winds or roof loads.

Exemptions to the regulation may still be required for existing facilities even with administrative and engineered improvements. In regard to consequences to the public, complete compliance with 10 CFR 70.61 using realistic assumptions should be the goal. Compliance with 10 CFR 70.61 regarding consequences to facility workers may require a request for an exemption once personnel protective equipment, emergency procedures, and worker training are taken into account. In the evaluation of a request for an exemption to the regulation, the expected operational life of the facility should also be factored into the determination of risk.

**Considerations for New Processes at Existing Facilities**

The design of new processes at existing facilities must address natural phenomena hazards in accordance with 10 CFR 70.64(a)(2), as well as the performance requirements of 10 CFR 70.61. Nevertheless, new processes at existing facilities may present the same problems in demonstrating compliance with 10 CFR 70.61 in regard to accident sequences initiated by natural phenomena as do existing facilities based on the design and/or siting of the original structures. In the case of new processes, the U.S. Nuclear Regulatory Commission staff should expect compliance with the performance requirements of 10 CFR 70.61 to the extent possible, given the existing facility design and location. New processes at existing facilities also must meet the requirements of 10 CFR 70.64(b), which requires defense in depth and a preference for engineered controls over administrative controls. However, the staff cannot require structural improvements, permanent flood barriers, and other engineered improvements that could be considered retrofits to be applied to existing structures. New structural features within existing structures to prevent breaches in containment in the event of natural phenomena hazards may be considered, however. An example might be a seismically designed vault to hold radioactive materials associated with a new process. In regard to new processes, engineered controls, where feasible, are preferred over administrative procedures that might otherwise be proposed for an existing process with a limited operational lifetime. Such engineered improvements may not be required for licensing but could be scheduled to replace administrative procedures or other long-term compensatory measures on a timely basis after the start of operations. The objective is to encourage engineered safety in new processes compared to equivalent existing processes, while recognizing the restraints of the existing structures and location. Although primarily aimed at reducing risk to the public, the emphasis on engineered safety may also be applied to worker consequences in a way consistent with the method accepted at other facilities.

## Regulatory Basis

The regulation in 10 CFR 70.61 specifies performance requirements associated with risks identified by an ISA.

For new facilities or new processes at existing facilities, 10 CFR 70.64 specifies requirements, including Baseline Design Criterion (a)(2), "Natural Phenomena Hazards."

## Technical Review Guidance

When examining the applicant's evaluation of the effects of natural phenomena on its facility, reviewers should recognize that estimates of "unlikely" and "highly unlikely" natural phenomena such as the PMF or SSE may not exist for the particular site. Hence, extrapolation and/or transposition of extreme event estimates made for other relatively nearby facilities (such as power reactor sites) should be allowed where feasible and technically justifiable. In addition, sophisticated probabilistic tools such as Bayesian analysis or Monte Carlo sampling methods need not be employed to improve the estimate of likelihoods of natural phenomena event sequences unless desired by the applicant (or licensee). For the purpose of determining appropriate values of extreme events, deterministic events such as the PMF or SSE can be used in place of purely probabilistically determined "highly unlikely" events and may be preferable, depending on the quality of historical data. Where extreme events need to be coupled with other probability-driven mechanisms, such as the release fraction or transport pathway, already low likelihood combinations do not have to be made even less likely by the use of conservative parameters.

For existing facilities, due credit should be given to analysis assumptions and administrative controls, emergency procedures, and active engineered controls that do not change the design bases of the facility structures to natural phenomena. If the ISA and ISA Summary demonstrate that the existing facility is near compliance (within an order of magnitude of a likelihood threshold or within 50 percent of meeting a consequence threshold, but not both), an exemption to the regulation may be considered.

The annex to this appendix presents an example of an evaluation for an amendment request.

## Recommendation

This guidance should be used to supplement Chapter 3 of this SRP.

## References

American National Standards Institute/American Nuclear Society, "Determining Design Basis Flooding at Power Reactor Sites," ANS-2.8, July 1992.

*U.S. Code of Federal Regulations*, Chapter I, Title 10, "Energy," Part 70, "Domestic Licensing of Special Nuclear Material."

U.S. Department of Energy, "Natural Phenomena Hazards Design and Evaluation Criteria for Department of Energy Facilities," DOE Standard-1020-2002, 2002.

U.S. Nuclear Regulatory Commission, "Domestic Licensing of Special Nuclear Material; Possession of a Critical Mass of Special Nuclear Material," *Federal Register*, Vol. 65, No. 181, pp. 56211–562331, September 18, 2000.

U.S. Nuclear Regulatory Commission, "Procedural and Submittal Guidance for the Individual Plant Examination of External Events (IPEEE) for Severe Accident Vulnerabilities," NUREG-1407, June 1991.

U.S. Nuclear Regulatory Commission, "Design Basis Flooding for Nuclear Power Plants," Regulatory Guide 1.59, Revision 2, August 1997.

# ANNEX TO APPENDIX D

# EXAMPLE OF NATURAL PHENOMENA HAZARD REVIEW FOR COMPLIANCE WITH 10 CFR 70.61

This example review is for an amendment to authorize operations in a blended low-enriched uranium oxide conversion building (OCB). The site is located near a river and is just above the 100-year flood plain of a nearby creek. The effluent process building was also part of the amendment but was not evaluated because the quantities of radioactive material or hazardous chemicals (that come under U.S. Nuclear Regulatory Commission (NRC) regulation) that it contained are not considered sufficient to exceed the consequence threshold for "unlikely" events given in Title 10 of the *Code of Federal Regulations* (10 CFR) 70.61, "Performance Requirements."

## Seismic Evaluation

The OCB is of reinforced concrete construction and is built to seismic criteria in the Standard Building Code (SBC-1999), which is equivalent to being designed for an earthquake with a probability of exceedance of approximately $4 \times 10^{-4}$ per year. Using Appendix C to U.S. Department of Energy (DOE) Standard-1020-2002, "Natural Phenomena Hazards Design and Evaluation Criteria for Department of Energy Facilities," the NRC staff determined the risk reduction factor to be 4, which gives the structure a likelihood of significant damage from an earthquake of $10^{-4}$ per year or less. Hence, the collapse or loss of building integrity from an earthquake may be considered to be "highly unlikely," as the probabilistic value of "highly unlikely" indicated by the applicant was a probability of exceedance of $10^{-4}$ to $10^{-5}$ per year. Within the building, the material at risk consists of low-enriched uranyl nitrate liquid, ammonium diuranate slurry, and uranium dioxide powder. All of these materials are expected to be within containers, and spillage during a seismic event is expected to be minimal. Since the building is expected to retain its integrity, the leak path factor will be relatively minor even without dynamic confinement from the ventilation system. Facility workers are expected to take actions to limit personal intake of radionuclides. The staff concludes that the OCB complies with the performance requirements of 10 CFR 70.61 with regard to seismic events.

## High-Winds Evaluation

The OCB structure is also designed for wind loads in accordance with SBC-1999, and the probability of a tornado impacting the facility is less than $10^{-5}$ per year. Therefore, the facility needs to be evaluated only in regard to the effects of wind loads and missiles, but not for tornadoes. The NRC staff considers the reinforced concrete exterior walls of the OCB to be adequate to withstand high wind velocities, as well as the missiles (from DOE Standard-1020-2002) that should be assumed for such events. The staff considers a collapse of building walls because of wind forces such that radioactive material would escape to be "highly unlikely." In addition, the meteorological conditions likely to result in severe winds may be forecast in advance and protective measures taken. The staff concludes that the OCB complies with the performance requirements of 10 CFR 70.61 with regard to wind events.

## Flooding Evaluation

The lowest floor in the OCB is 15 feet above the 100-year flood from an adjacent creek. From a review of the topography of the site area, it appears that flooding of the site could occur, most likely from flooding of the nearby river with coincident flooding of the adjacent creek, which could back up through the railroad culvert. This event is expected to have warning time and may overtop the railroad embankment to the north of the facility and flood parts of the nearby town. However, the facility is sufficiently removed from the main channel of the river that flood-induced scouring and erosion would not be expected. In addition, the hydrostatic loading from the flood on the exterior walls of the OCB would not be expected to cause collapse. The primary concern is inundation, which could float unsecured containers within the OCB but not remove them from the facility. A criticality event cannot be excluded, but it could occur only in the flooded and, therefore, evacuated section of the plant and would not affect facility workers. In addition, the warning time would allow the movement of material to reduce the likelihood of a flood-induced criticality. The staff concludes that the OCB complies with the performance requirements of 10 CFR 70.61 with regard to flooding.

## REFERENCES

*U.S. Code of Federal Regulations*, Chapter I, Title 10, "Energy," Part 70, "Domestic Licensing of Special Nuclear Material."

U.S. Department of Energy, "Natural Phenomena Hazards Design and Evaluation Criteria for Department of Energy Facilities," DOE Standard-1020-2002, 2002.

# APPENDIX E

# HUMAN FACTORS ENGINEERING FOR PERSONNEL ACTIVITIES

The purpose of this review is to establish that human factors engineering (HFE) is applied to personnel activities identified as safety significant, consistent with the findings of the integrated safety analysis (ISA), and the determination of whether an item relied on for safety (IROFS) has special or unique safety significance. A graded approach commensurate with the complexity and integration and operation of the control systems is appropriate. The application of HFE to personnel activities ensures that the potential for human error in the facility operations was addressed during the design of the facility by facilitating correct, and inhibiting wrong, decisions by personnel and by providing a means for detecting and correcting or compensating for error.

Title 10 of the *Code of Federal Regulations* (10 CFR) 70.61(e) requires a safety program to ensure that each IROFS will be available and reliable to perform its intended function when needed. Therefore, the applicant should identify those "personnel activities[1]" that are considered IROFS and personnel activities that support safety (e.g., maintenance). An HFE review should be performed to demonstrate compliance with 10 CFR 70.61(e). Also, the applicant should demonstrate how personnel activities will enhance safety by reducing challenges to IROFS, as required in 10 CFR 70.64(b)(2).

A human factors specialist and an ISA reviewer should conduct the human factors review. The review should also be coordinated with the reviewers of other technical areas and the reviewer of management measures, as necessary.

## AREAS OF REVIEW AND ACCEPTANCE CRITERIA

Some facilities rely heavily on automated systems employing advanced digital instrumentation and control technology. These systems may be complex, with potential negative impacts on human performance activities in both operations and maintenance. The scope of review for the HFE for personnel activities should be consistent with the results of the ISA and include the following, as appropriate[2]:

A.  Identification of Personnel Activities—The applicant should appropriately identify the personnel activities such that the reviewer can understand the actions, the human-systems interfaces (HSIs) involved, and the consequences.

B.  HFE Design Review Planning—The applicant's approach for planning HFE design review should include the following:

   i.   Identification of appropriate goals and scope to ensure that HFE practices and guidelines are implemented during design, construction, and operation of the facility.

---

[1]  For the purposes of this chapter, the phrase "personnel activities" represents personnel activities identified as IROFS and personnel activities that support safety, such as maintenance.

[2]  All nine areas of review (A through I) may not be necessary for a specific application. Areas of review should be based on the applicant's provisions to address personnel activities consistent with the ISA findings, the similarity of the associated HFE issues for similar type plants, and the determination of whether an IROFS has special or unique safety significance.

ii.　Implementation by an HFE team that has the appropriate composition, experience, and organizational authority to ensure that HFE is considered in the design of HSI for personnel activities. The HFE team's responsibilities include ensuring the proper development, execution, oversight, and documentation of the HFE function. Depending on the identification of personnel activities, it may be appropriate for the HFE team to consist of a single individual.

iii.　An HFE team that attains the HFE goals and scope through established processes and procedures and that tracks HFE issues. The HFE function that should ensure that all aspects of the personnel activities including the HSI are developed, designed, and evaluated on the basis of a structured approach using HFE.

C.　Operating Experience Review (OER)—To the extent possible, the applicant should identify safety-related HFE events or potential events in existing facilities that are similar to the proposed facility. The applicant should do all of the following:

i.　Review the HFE-related events or potential events for relevance.

ii.　Analyze the HSI technology employed for the relevant HFE events or potential events.

iii.　Conduct (or review existing) operator interviews and surveys on the HSI technology for the relevant HFE events or potential events.

D.　Functional Allocation Analysis and Task Analysis

i.　The functional allocation analysis should be based on the OER. Personnel activities should be functionally allocated to take advantage of human strengths and to avoid demands that are not compatible with human capabilities.

ii.　The task analysis should include the task analysis scope, identification, and analysis of critical tasks; detailed description of personnel demands (e.g., input, processing, and output); iterative nature of the analysis; and incorporation of job design issues. The task analysis should address each operating mode for each personnel activity (e.g., startup, normal operations, emergency operations, and shutdown). The task analysis results support the functional allocation.

E.　HSI Design, Inventory, and Characterization—The HSI design should incorporate the functional allocation analysis and task analysis into the detailed design of safety-significant HSI components (e.g., alarms, displays, controls, and operator aids) through the systematic application of HFE. The HSI design should include the overall work environment, the work space layout (e.g., control room and remote shutdown facility layouts), the control panel and console design, the control and display device layout, and information and control interface design details. The HSI design process should ensure the application of HFE to the HSI required to perform personnel activities. The HSI design process should exclude the development of extraneous controls and displays. The HSI design documentation should include a complete HSI inventory and the basis for the HSI characterization.

F.  Staffing—Staffing should be based on a review of the number and qualifications of personnel for each personnel activity during all plant operating conditions. The applicant should conduct this review in a systematic manner that incorporates the functional allocation and task analysis results.

Categories of personnel should be based on the types of personnel activities. Staffing considerations should include issues identified in the OER, functional allocation, HSI design, procedure development, and verification and validation (V&V).

G.  Procedure Development—The applicant's procedure development for personnel activities should incorporate HFE principles and criteria, along with all other design requirements, to develop procedures that are technically accurate, comprehensive, explicit, easy to use, and validated consistent with the acceptance criteria in this Standard Review Plan. Because procedures are considered an essential component of the HSI design, they should be derived from the same design process and analyses as the other components of the HSI (for example, displays, controls, operator aids) and subject to the same evaluation processes. Procedures to support the personnel activity may include generic technical guidance, plant and system operations, abnormal and emergency operations, tests (for example, preoperational, startup, and surveillance), and alarm response.

H.  Training Program Development—The applicant's training program development should address all personnel activities. The training program development indicates how the knowledge and skill requirements of personnel will be evaluated, how the training program development will be coordinated with the other activities of the HFE design process, and how the training program will be implemented in an effective manner consistent with human factors principles and practices.

The training program development should address the areas of review and acceptance criteria described in Chapter 11 of this SRP and should result in a training program that provides personnel with qualifications commensurate with their activities.

I.  Verification & Validation—V&V confirms that the design incorporates HFE to HSI in a manner that enables the successful completion of personnel activities. The V&V should be applied to personnel activities (see item A) and HSI design (see item E). The V&V process should consist of the following:

    i.  HSI task support verification: HSI components should be appropriately provided for personnel activities through HSI task support verification. The verification should show that each HSI has identified the task analysis (see item D(ii)) and that the HSI design (see item E) is appropriately provided, yet minimizes the incorporation of information, displays, controls, and embellishments that unnecessarily complicate personnel activities.

    ii. HFE design verification: The HFE design verification should show that each HSI identified for a personnel activity has incorporated HFE into the design. Deviations from accepted HFE principles and guidelines should be justified or documented for resolution or correction. If HFE design verification does not address all HSI components, then an alternative multidimensional sampling methodology should be used to ensure comprehensive consideration of the safety significance of HSI components. The sample size should be sufficient to identify a range of significant safety issues.

iii.     Integrated system validation: The applicant should conduct a performance-based evaluation of the integrated design to ensure that the HFE/HSI supports safe operation of the plant. Integrated system validation is performed after HFE problems identified in HFE design activities are resolved or corrected because these may negatively affect performance and, therefore, validation results. Validation is performed by evaluating personnel activities using appropriate measurement tools. All personnel activities should be tested and found to be adequately supported in the design, including personnel activities outside the control room.

iv.     Human factors issue resolution verification: The applicant should verify that HFE issues identified during the design process were addressed and resolved. Issue resolution verification should be documented in the HFE issue tracking system established by the HFE team (see item B). Issues that cannot be resolved until the HSI design is constructed, installed, and tested should be identified and incorporated into the final HFE/HSI design verification.

v.     Final HFE/HSI design verification: The applicant should commit to performing a final HFE/HSI design verification if the applicant cannot demonstrate that it has fully evaluated the actual installation of the final HSI design in the plant through the V&V activities described above. Final HFE/HSI design verification should demonstrate that in-plant HFE design implementation conforms to the HFE design (see item E) as modified by V&V activities. V&V activities should be performed in the order listed above, as necessary. However, the applicant may find that it is necessary to repeat the activities in order to address design corrections and modifications that occur during V&V.

## REFERENCES

*U.S. Code of Federal Regulations*, Chapter I, Title 10, "Energy," Part 70, "Domestic Licensing of Special Nuclear Material."

U.S. Nuclear Regulatory Commission, "Standard Review Plan for the Review of a License Application for a Mixed Oxide Fuel Fabrication Facility," NUREG-1718, August 2000.

# 4. RADIATION PROTECTION

## 4.1 Purpose of Review

The purpose of this review is to determine whether the applicant's radiation protection program is adequate to protect the radiological health and safety of workers and to comply with the regulatory requirements in Title 10 of the *Code of Federal Regulations* (10 CFR) Part 19, "Notices, Instructions and Reports to Workers: Inspection and Investigations"; 10 CFR Part 20, "Standards for Protection Against Radiation"; and 10 CFR Part 70, "Domestic Licensing of Special Nuclear Material."

The content and level of detail in this chapter are generally greater than in other chapters because this chapter provides acceptance criteria for evaluating compliance with 10 CFR Part 20, which has very specific requirements. The applicant should also incorporate, and the U.S. Nuclear Regulatory Commission (NRC) reviewer should consider, insights gained from the conduct of the integrated safety analysis (ISA) and information contained in the ISA Summary in developing and reviewing the acceptability of the applicant's radiation protection program. In addition, the reviewer should evaluate the adequacy of the ISA Summary with respect to ensuring that the application meets the radiation exposure performance criteria of 10 CFR 70.61(b) and (c). Chapter 9 of this Standard Review Plan (SRP), which discusses environmental protection, contains the review procedures and acceptance criteria for the applicant's program for protecting members of the public and controlling effluent releases.

## 4.2 Responsibility for Review

Primary:        Health Physicist

Secondary:    Licensing Project Manager, Environmental Reviewer

Supporting:   Fuel Cycle Facility Inspector

## 4.3 Areas of Review

The radiation protection program must address the occupational radiation protection measures in 10 CFR Parts 19, 20, and 70. Specifically, licensees must develop, document, and implement a radiation protection program in accordance with 10 CFR 20.1101, "Radiation Protection Programs." Additionally, 10 CFR 20.2102, "Records of Radiation Protection Programs," requires licensees to keep records of the radiation protection program, including a description of the program components, audits, and other aspects of program implementation. The reviewer should also refer to the ISA Summary to identify those facility operations analyzed in the ISA that have radiological consequences and the items relied on for safety (IROFS), as well as the management measures implemented to prevent or mitigate such radiological risks. The ISA review should include a judgment as to the comprehensiveness of evaluations performed by the licensee.

The staff will review an applicant's commitments regarding the following components of the radiation protection program:

- Establish, maintain, and implement a radiation protection program.

- Keep occupational exposures to radiation as low as reasonably achievable (ALARA).

- Appoint radiological protection staff who are suitably qualified and trained in radiation protection procedures.

- Prepare written radiation protection procedures and radiation work permits (RWPs).

- Train employees in radiation protection, including the health protection problems associated with exposure to radiation, precautions and procedures to minimize exposure, and the purposes and functions of protective devices employed.

- Design and implement programs to control airborne concentrations of radioactive material by using ventilation systems, containment systems, and respirators.

- Conduct radiation surveys and monitoring programs to document radiation levels, concentrations of radioactive materials in the facility, and occupational exposures to radiation by workers.

- Evaluate the radiological risks from accidents occurring during operations; identify IROFS that limit high and intermediate consequences, consistent with regulatory performance criteria; and have appropriate management measures in place to ensure that identified IROFS are available and reliable.

- Maintain additional programs, including (1) a records maintenance program, (2) a corrective action program, and (3) a program for reporting to the NRC in accordance with requirements in 10 CFR Part 20 and 10 CFR Part 70.

Review Interfaces

In addition to Chapter 4 of the application, the reviewer should examine information in the following other areas to ensure that it is consistent with the information in Chapter 4:

- emergency plan applicable to radiation protection under SRP Chapter 8

- safety program, ISA commitments, and ISA documentation applicable to radiation protection under SRP Chapter 3

- environmental and effluent monitoring, as well as any effluent controls applicable to radiation protection under SRP Chapter 9

## 4.4 Acceptance Criteria

### 4.4.1 Commitment to Radiation Protection Program Implementation

#### 4.4.1.1 Regulatory Requirements

Regulations in 10 CFR 20.1101 apply to the establishment of a radiation protection program.

#### 4.4.1.2 Regulatory Guidance

The NRC regulatory guide applicable to the commitment to design and implement a radiation protection program is Regulatory Guide 8.2, "Guide for Administrative Practice in Radiation Monitoring," issued February 1973.

#### 4.4.1.3 Regulatory Acceptance Criteria

The applicant's radiation protection program is acceptable if the license application provides data and information that meet each of the following commitments:

- Design and implement a radiation protection program that meets the regulatory requirements of 10 CFR 20.1101.

- Outline the radiation protection program structure and define the responsibilities of key program personnel.

- Staff the radiation protection program with suitably trained people, provide sufficient resources, and implement the program.

- Commit to the independence of the radiation protection function from the facility's operations.

- Review, at least annually, the content and implementation of the radiation protection program, as required by 10 CFR 20.1101(c). The review should consider facility changes, new technologies, and other process enhancements that could improve the effectiveness of the overall program.

### 4.4.2 Commitment to an ALARA Program

#### 4.4.2.1 Regulatory Requirements

Regulations in 10 CFR 20.1101 apply to the ALARA program.

#### 4.4.2.2 Regulatory Guidance

The following NRC regulatory guides are applicable to the ALARA program:

- Regulatory Guide 8.2

- Regulatory Guide 8.10, "Operating Philosophy for Maintaining Occupational Radiation Exposures as Low as Is Reasonably Achievable," Revision 1-R, May 1977

- Regulatory Guide 8.13, "Instructions Concerning Prenatal Radiation Exposure," Revision 3, June 1999

- Regulatory Guide 8.15, "Acceptable Programs for Respiratory Protection," Revision 1, May 1977

- Regulatory Guide 8.29, "Instructions Concerning Risks from Occupational Radiation Exposure," February 1996

- Regulatory Guide 4.21, "Minimization of Contamination and Radioactive Waste Generation: Life Cycle Planning," June 2008

### 4.4.2.3 Regulatory Acceptance Criteria

The applicant's ALARA program is acceptable if the license application provides data and information that meet each of the following commitments:

- Establish a written, comprehensive, and effective ALARA program.

- Prepare policies and procedures to ensure that occupational radiation exposures are maintained ALARA and that such exposures are consistent with the requirements of 10 CFR 20.1101.

- Outline specific ALARA program goals, establish an ALARA program organization and structure, and include written procedures for its implementation in the plant design and operations.

- Establish an ALARA committee, or equivalent organization, with sufficient staff, resources, and clear responsibilities to ensure that the occupational radiation exposure does not exceed the dose limits of 10 CFR Part 20 under normal operations.[1]

- Use the ALARA program as a mechanism to facilitate interaction between radiation protection and operations personnel.

- Regularly review and revise, when appropriate, the ALARA program goals and objectives and incorporate, when appropriate, new approaches, technologies, operating procedures, or changes that could reduce potential radiation exposures at a reasonable cost.

### 4.4.3 Organization and Personnel Qualifications

---

[1] The ALARA committee should meet at least annually, and the membership should include areas such as management, radiation protection, environmental safety, industrial safety, and production. The committee's review of the ALARA program should include an evaluation of the results of audits made by the radiation protection organization, reports of radiation levels in the facility, contamination levels, employee exposures, and effluent releases. The review should determine if there are any upward trends in personnel exposure for identified categories of workers and types of operations. The review should identify any upward trends in effluent releases and contamination levels. Finally, the review should determine if exposures, releases, and contamination levels are in accordance with the ALARA concept. The ALARA committee should document its recommendations and track them to completion.

### 4.4.3.1 Regulatory Requirements

Regulations in 10 CFR 70.22, "Contents of Applications," apply to the organization and qualifications of the radiological protection staff.

### 4.4.3.2 Regulatory Guidance

The following are the NRC regulatory guides applicable to the organization and personnel qualifications of radiation protection program staff:

- Regulatory Guide 8.2

- Regulatory Guide 8.10

### 4.4.3.3 Regulatory Acceptance Criteria

The applicant's commitment to organize and staff a radiation protection program is acceptable if the license application provides data and information that meet each of the following commitments:

- Appoint radiation protection personnel and identify their authority and responsibilities for implementing the radiation protection program functions.

- Establish clear organizational relationships among the individual positions responsible for the radiation protection program and other line managers.

- Appoint a suitably educated, experienced, and trained radiation protection program director (typically referred to as the radiation safety officer) who (1) has direct access to the plant manager, (2) is skilled in the interpretation of data and regulations pertinent to radiation protection, (3) is familiar with the operation of the facility and radiation protection concerns of the site, (4) participates as a resource in radiation safety management decisions, and (5) will be responsible for establishing and implementing the radiation protection program.

- Describe the minimum education, experience, and training requirements for the radiation protection program director and staff.

## 4.4.4 Commitment to Written Procedures

### 4.4.4.1 Regulatory Requirements

The regulations in 10 CFR 70.22(a)(8) apply to radiation protection procedures and RWPs.

### 4.4.4.2 Regulatory Guidance

The regulatory guidance applicable to procedures and RWPs appears in Regulatory Guide 8.10, Revision 1-R.

### 4.4.4.3 Regulatory Acceptance Criteria

The applicant's commitment to prepare written radiation protection procedures and RWPs is acceptable if the license application provides data and information that meet each of the following commitments:

- Prepare written, approved procedures to carry out activities related to the radiation protection program.

- Establish a process for procedure generation or modification, authorization, distribution, and training, such that changes in technology or practices are communicated effectively and in a timely manner. Review and revise procedures, as necessary, to incorporate any facility or operational changes, including changes in the ISA. The radiation safety officer, or an individual who has the qualifications of the radiation safety officer, should approve all procedures related to radiation protection.

- Specify written, approved RWPs for activities involving licensed material that are not covered by written radiation protection procedures. RWPs should define the authorized activities, the level of approval required (a radiation specialist, as a minimum), information requirements, period of validity, expiration and termination times, and recordkeeping requirements.

### 4.4.5 Radiation Safety Training

The SRP addresses an applicant's commitments to employee training in several places. This chapter addresses corporate radiation protection training programs, and Chapter 11 discusses training that serves as a management measure for ensuring that an administrative control IROFS is available and reliable when required.

### 4.4.5.1 Regulatory Requirements

The following regulations apply to the radiation safety training program:

- 10 CFR 19.12, "Instructions to Workers"

- 10 CFR 20.2110, "Form of Records"

### 4.4.5.2 Regulatory Guidance

The following NRC regulatory guides; reports of the National Council on Radiation Protection (NCRP); and standards of the American National Standards Institute (ANSI)/Health Physics Society (HPS) and the American Society for Testing and Materials pertain to radiation protection training:

- Regulatory Guide 8.10

- Regulatory Guide 8.13

- Regulatory Guide 8.29

- NCRP Report No. 134, "Operational Radiation Safety Training," 2000

- ASTM E1168-95, "Radiological Protection Training for Nuclear Facility Workers," reapproved in 2008

- ANSI/HPS N13.36, "Ionizing Radiation Safety Training for Workers," October 30, 2001

*4.4.5.3 Regulatory Acceptance Criteria*

The applicant's commitment to train its employees in radiation protection is acceptable if the license application provides data and information that meet each of the following commitments:

- Design and implement an employee radiation protection training program that complies with the requirements of 10 CFR Parts 19 and 20.

- Provide training to all personnel and visitors entering restricted areas that is commensurate with the health risk to which they may be exposed, or provide escorts who have received the appropriate training.

- Provide a level of training commensurate with the potential radiological health risks associated with that employee's work responsibilities.

- Conduct refresher training at least every 3 years that will accurately address changes in policies, procedures, requirements, and the facility ISA.

- Incorporate into the radiation protection training program the provisions in 10 CFR 19.12 and additional relevant topics, such as the following (the asterisk denotes those topics with a basis in 10 CFR 19.12):

  - correct handling of radioactive materials

  - the storage, transfer, or use of radiation or radioactive material as relevant to the individual's activities*

  - minimization of exposures to radiation or radioactive materials*

  - access and egress controls and escort procedures

  - radiation safety principles, policies, and procedures*

  - monitoring for internal and external exposures

  - radiation exposure reports available to workers*

  - monitoring instruments

  - contamination control procedures, including protective clothing and equipment*

  - ALARA and exposure limits*

- radiation hazards and health risks*

- emergency response*

- responsibility to report promptly any condition that may lead to, or cause, a violation of regulations and licenses or create unnecessary exposure*

- Review and evaluate the accuracy, effectiveness, and adequacy of the radiation protection training program curriculum and instructors, as applicable, at least every 3 years.

## 4.4.6 Ventilation and Respiratory Protection Programs

### 4.4.6.1 Regulatory Requirements

Regulations in Subpart H, "Respiratory Protection and Controls to Restrict Internal Exposure in Restricted Areas," to 10 CFR Part 20 apply to the ventilation and respiratory protection programs.

### 4.4.6.2 Regulatory Guidance

The following NRC regulatory guides, ANSI standards, and other publications apply to the design of the ventilation and respiratory protection programs:

- Regulatory Guide 8.24, "Health Physics Surveys During Enriched Uranium-235 Processing and Fuel Fabrication," Revision 1, October 1979

- Regulatory Guide 8.15

- American Conference of Governmental Industrial Hygienists (ACGIH) 2095, "Industrial Ventilation: A Manual of Recommended Practice for Design," 2007

- Energy Research and Development Administration (ERDA) 76-21, "Nuclear Air Cleaning Handbook," by C.A. Burchsted, A.B. Fuller, and J.E. Kahn, March 31, 1976

- ANSI Z88.2-1992, "Practices for Respiratory Protection"

### 4.4.6.3 Regulatory Acceptance Criteria

The applicant's commitment to have ventilation and respiratory protection programs is acceptable if the license application provides data and information that meet each of the following commitments:

- Install appropriately sized ventilation and containment systems in areas of the plant identified as having potential airborne concentrations of radionuclides that could exceed the occupational derived air concentration values specified in 10 CFR Part 20, Appendix B, "Annual Limits on Intake (ALIs) and Derived Air Concentrations (DACs) of

Radionuclides for Occupational Exposure; Effluent Concentrations; Concentrations for Release to Sewerage," during normal operations.

- Describe management measures, including preventive and corrective maintenance and performance testing, to ensure that the ventilation and containment systems operate when required and are within their design specifications.

- Describe the criteria for the ventilation and containment systems, including minimum flow velocity at openings in these systems, maximum differential pressure across filters, and types of filters to be used.

- Describe the frequency and types of tests to measure the performance of ventilation and containment systems, the acceptance criteria, and the actions to be taken when the acceptance criteria are not satisfied.

- Establish a respiratory protection program that meets the requirements of 10 CFR Part 20, Subpart H.

- Prepare written procedures for the selection, fitting, issuance, maintenance, testing, training of personnel, monitoring, and recordkeeping for individual respiratory protection equipment and for specifying when such equipment is to be used.

- Revise the written procedures for the use of individual respiratory protection equipment, as applicable, when making changes to processing, facility, or equipment.

- Maintain records of the respiratory protection program, including training in respirator use and maintenance.

### 4.4.7 Radiation Surveys and Monitoring Programs

Radiation surveys are conducted for two purposes: (1) to ascertain radiation levels, concentrations of radioactive material, and potential radiological hazards that could be present in the facility and (2) to detect releases of radioactive material from plant equipment and operations. Radiation surveys will focus on those areas of the plant necessary to show compliance with the dose limits and monitoring requirements of 10 CFR Part 20, Subpart C, "Occupational Dose Limits"; Subpart D, "Radiation Dose Limits for Individual Members of the Public"; and Subpart F, "Surveys and Monitoring".

Measurements of airborne radioactive material and bioassays are used to determine internal occupational exposures to radiation. When combined with external occupational exposure data, the dose of record can be compared against the dose limits specified in 10 CFR Part 20, Subpart C.

*4.4.7.1 Regulatory Requirements*

The following NRC regulations in 10 CFR Part 20 apply to radiation surveys and monitoring programs:

- Subpart F

- Subpart C
- Subpart L, "Records"
- Subpart M, "Reports"

*4.4.7.2 Regulatory Guidance*

The following NRC regulatory guides, NUREGs, and ANSI standards are applicable to radiation surveys and monitoring programs:

- Regulatory Guide 8.2

- Regulatory Guide 8.4, "Direct-Reading and Indirect-Reading Pocket Dosimeters," February 1973

- Regulatory Guide 8.7, "Instructions for Recording and Reporting Occupational Radiation Exposure Data," Revision 2, November 2005

- Regulatory Guide 8.9, "Acceptable Concepts, Models, Equations, and Assumptions for a Bioassay Program," Revision 1, July 1993

- Regulatory Guide 8.24

- Regulatory Guide 8.25, "Air Sampling in the Workplace," Revision 1, June 1992

- Regulatory Guide 8.34, "Monitoring Criteria and Methods To Calculate Occupational Radiation Doses," July 1992

- NUREG-1400, "Air Sampling in the Workplace," September 1993

- ANSI N13.1-1999, "Sampling and Monitoring Releases of Airborne Radioactive Substances from the Stacks and Ducts of Nuclear Facilities"

- ANSI N328-1978, "Radiation Protection Instrumentation Test and Calibration"

- ANSI N13.11-2001, "Personnel Dosimetry Performance—Criteria for Testing"

- ANSI N13.15-1985, "Radiation Detectors—Personnel Thermoluminescence Dosimetry Systems—Performance"

- ANSI.N13.22-1995, "Bioassay Program for Uranium"

- ANSI N13.27-1981, "Performance Requirements for Pocket-Sized Alarm Dosimeters and Alarm Ratemeters"

- ANSI N13.30-1996, "Performance Criteria for Radiobioassay"

- ANSI N13.6-1999, "Practice for Occupational Radiation Exposure Records Systems"

*4.4.7.3 Regulatory Acceptance Criteria*

The applicant's commitment to implement radiation surveys and monitoring programs is acceptable if the license application provides data and information that meet each of the following commitments:

- Provide radiation survey and monitoring programs that are consistent with the requirements of 10 CFR Part 20, Subpart F.

- Prepare written procedures for the radiation survey and monitoring programs that include an outline of the program objectives, sampling procedures, data analysis methods, types of equipment and instrumentation, frequency of measurements, recordkeeping and reporting requirements, and actions to be taken when measurements exceed occupational dose limits in 10 CFR Part 20 or administrative levels established by the applicant.

- Design and implement a personnel monitoring program for external occupational radiation exposures that outlines methods or procedures to do the following:

    - Identify the criteria for worker participation in the program.

    - Identify the types of radiation to be monitored.

    - Specify how exposures will be measured, assessed, and recorded.

    - Identify the type and sensitivity of personal dosimeters to be used, when they will be used, and how they will be processed and evaluated.

    - Identify the plant's administrative exposure levels or the levels at which actions are taken to investigate the cause of exposures exceeding these levels.

- Design and implement a personnel monitoring program for internal occupational radiation exposures, based on the requirements of 10 CFR 20.1201, "Occupational Dose Limits for Adults"; 10 CFR 20.1204, "Determination of Internal Exposure"; and 10 CFR 20.1502(b), that outlines methods or procedures to do the following:

    - Identify the criteria for worker participation in the program.

    - Identify the type of sampling to be used, the frequency of collection and measurement, and the minimum detection levels.

    - Specify how worker intakes will be measured, assessed, and recorded.

    - Specify how the data will be processed, evaluated, and interpreted.

    - Identify the plant's administrative exposure levels or the levels at which actions are taken to investigate the cause of exposures exceeding these levels.

- Comply with the requirements of 10 CFR 20.1202, "Compliance with Requirements for Summation of External and Internal Doses," for summation of external and internal occupational radiation exposures through the use of procedures such as those outlined in Regulatory Guides 8.7 or 8.34.

- Design and implement an air sampling program in areas of the plant identified as potential airborne radioactivity areas to conduct airflow studies and to calibrate and maintain the airborne sampling equipment in accordance with the manufacturers' recommendations.

- Implement additional procedures, as may be required by 10 CFR Part 20 and the ISA Summary, to control the concentration of airborne radioactive material (e.g., control of access, limitation of exposure times to licensed materials, and use of respiratory protection equipment).

- Conduct a contamination survey program in areas of the plant most likely to be radiologically contaminated (the program must include the types and frequencies of surveys for various areas of the plant and the action levels and actions to be taken when contamination levels are exceeded).

- Implement the facility's corrective action program when the results of personnel monitoring or contamination surveys exceed the applicant's administrative personnel contamination levels.

- Implement the facility's corrective action program when any incident results in airborne occupational exposures to radiation exceeding the facility's administrative limits, or the dose limits in Appendix B to 10 CFR Part 20 or in 10 CFR 70.61, "Performance Requirements."

- Use equipment and instrumentation with sufficient sensitivity for the type or types of radiation being measured and calibrate and maintain equipment and instrumentation in accordance with the manufacturers' recommendations or applicable ANSI standards.

- Establish policies to ensure that equipment and materials removed from restricted areas to unrestricted areas are not contaminated above the specified release levels in the NRC branch technical position entitled, "Guidelines for Decontamination of Facilities and Equipment Prior to Release for Unrestricted Use or Termination of Licenses for Byproduct, Source, or Special Nuclear Material," issued April 1993.

- Leak-test all sealed sources consistent with the applicable NRC branch technical positions issued in April 1993: (1) "License Condition for Leak-Testing Sealed Byproduct Material Sources," (2) "License Condition for Leak-Testing Sealed Plutonium Sources," (3) "License Condition for Plutonium Alpha Sources," (4) "License Condition for Leak-Testing Sealed Source Which Contains Alpha and/or Beta-Gamma Emitters," and (5) "License Condition for Leak-Testing Sealed Uranium Sources.

- Establish and implement an access control program that ensures that (1) signs, labels, and other access controls are properly posted and operative, (2) restricted areas are established to prevent the spread of contamination and are identified with appropriate signs, and (3) step-off pads, change facilities, protective clothing facilities, and personnel monitoring instruments are provided in sufficient quantities and locations.

- Establish a reporting program that is consistent with the requirements of 10 CFR Part 19 and 10 CFR Part 20.

### 4.4.8 Control of Radiological Risk Resulting from Accidents

In addition to participating in the integrated review of the ISA Summary performed in accordance with Chapter 3 of the SRP, the reviewer should also examine, in detail, the radiological exposure and release accident sequences provided in the ISA Summary to demonstrate compliance with 10 CFR 70.61. This review should include an evaluation of sequences involving radiological releases or exposures with respect to the initiators and their frequency, radiological consequences, and IROFS chosen to prevent or mitigate those consequences.

The reviewer should also identify and note any items or issues that should be inspected during an operational readiness review, if such will be performed. These items may include confirming that engineered controls meet performance specifications described in the ISA Summary and that administrative controls are implemented through procedures and operator training.

The reviewer should ensure that the emergency plan, if one is required, adequately addresses the licensee response to a release of radioactive materials or that the licensee gives a proper justification that precludes the development of an emergency plan.

Finally, the reviewer should be aware that accident sequences considered "not unlikely" in the ISA Summary are constricted, under the ALARA requirement in 10 CFR Part 20, to minimize exposure to personnel and the public.

#### 4.4.8.1 Regulatory Requirements

The following NRC regulations apply to the control of radiological risk from accidents:

- 10 CFR 70.22(i)(1) requires either an evaluation that the maximum dose to a member of the public resulting from a release of materials would not exceed 1 rem or 2 milligrams soluble uranium intake or the submission of an emergency plan for responding to the radiological hazards of a postulated accident.

- 10 CFR Part 70, Subpart H, contains requirements for performing ISAs, designating IROFS, and having management measures in place, both to ensure that IROFS are readily available and reliable and to provide facility change management and configuration control.

- 10 CFR 20.1101 states that licensees shall apply procedures and engineering controls to achieve exposures to workers and the public that are ALARA.

- 10 CFR 20.1406, "Minimization of Contamination," states that licensees shall design and develop procedures for operation that will minimize contamination of the facility and the environment, facilitate eventual decommissioning, and minimize the generation of radioactive waste.

- 10 CFR Part 20, Subpart H, discusses controls to restrict internal exposures.

*4.4.8.2 Regulatory Guidance*

The following guidance applies to the control of radiological risk resulting from accidents:

- NUREG-1513, "Integrated Safety Analysis Guidance Document," May 2001

- NUREG/CR-6410, "Nuclear Fuel Cycle Accident Analysis Handbook," March 1998

- Regulatory Guide 4.21

*4.4.8.3 Acceptance Criteria*

The reviewer should consider the factors listed below in determining the acceptability of the applicant's descriptions of radiological exposure or release accident sequences.

- Accident sequences should be sufficiently described and detailed to allow an understanding of the radiological hazards (e.g., radioactive materials at risk) and the release mechanism.

- The applicant should provide adequate descriptions of the radiological consequences (i.e., exposure estimates) identified in the ISA Summary. The reviewer should verify that exposures are reasonable, based on the sequence description and the radioactive materials involved, and use a methodology consistent with regulatory guidance (10 CFR 70.61).

- The applicant should justify the likelihood of the initiating event, its prevention, or mitigation of the results of an accident sequence with high or intermediate consequences, if credited in a questionable or nonconservative manner. If controls are relied on to reduce the likelihood or severity of a high- or intermediate-consequence accident sequence, they should be identified as IROFS (10 CFR 70.61).

- Analyses that the applicant has performed as part of the ISA process should be referenced or identified for potential further review (vertical slice) by the NRC staff (10 CFR 70.61).

- The application should demonstrate the management measures proposed to ensure that IROFS are available and reliable, when required, by briefly describing both of the following:

    – procedures to ensure the reliable operation of engineered controls (e.g., inspection and testing procedures and frequencies, calibration programs, functional tests, corrective and preventive maintenance programs, and criteria for acceptable test results) (10 CFR 70.62(d))

    – procedures to ensure that administrative controls will be correctly implemented when required (e.g., employee training and qualification in operating procedures, refresher training, safe work practices, development of standard operating procedures, and training program evaluations) (10 CFR 70.62(d))

- The application shall include either of the following:

    - an evaluation that demonstrates that public exposures resulting from offsite releases of material are less than 1 rem or 2 milligrams soluble uranium intake (10 CFR 70.22(i)(1)(i))

    - an emergency plan that includes sufficient detail for responding appropriately to an offsite release of radioactive materials (10 CFR 70.22(i)(1)(ii))

## 4.4.9 Additional Program Commitments

### 4.4.9.1 Regulatory Requirements

The following 10 CFR Part 20 regulations are applicable to the additional program commitments:

- Subpart L
- Subpart M
- 10 CFR 70.74, "Additional Reporting Requirements"

### 4.4.9.2 Regulatory Guidance

No NRC regulatory guidelines apply to these additional program commitments.

### 4.4.9.3 Acceptance Criteria

The applicant's commitment to implement additional program features is acceptable if the license application provides data and information that meet each of the following commitments:

- Maintain records of the radiation protection program (including program provisions, audits, and reviews of the program content and implementation), radiation survey results (air sampling, bioassays, external exposure data from monitoring of individuals, internal intakes of radioactive material), results of corrective action program referrals, RWPs, and planned special exposures.

- Establish a program to report to the NRC, within the time specified in 10 CFR 20.2202, "Notification of Incidents," and 10 CFR 70.74, any event that results in an occupational exposure to radiation exceeding the dose limits in 10 CFR Part 20.

- Prepare and submit to the NRC an annual report of the results of individual monitoring, as required by 10 CFR 20.2206(b).

- Refer to the facility's corrective action program any incident that results in an occupational exposure to radiation that exceeds the dose limits in Appendix B to 10 CFR Part 20 or in 10 CFR 70.74, and report to the NRC both the corrective action taken (or planned) to protect against a recurrence and the proposed schedule to achieve compliance with the applicable license condition or conditions.

## 4.5  Review Procedures

### 4.5.1  Acceptance Review

The primary reviewer should evaluate the license application to determine whether it addresses the areas of review discussed in Section 4.3.  If the reviewer identifies significant deficiencies, the agency should ask the applicant to submit additional material before starting the safety evaluation.

### 4.5.2  Safety Evaluation

The primary reviewer will perform a safety evaluation with respect to the acceptance criteria in Section 4.4.  For existing facilities, the reviewer will consult with the cognizant NRC inspector for radiation protection to identify and resolve any issues of concern related to the licensing review.  The primary reviewer will prepare a safety evaluation report (SER) on the licensing action for the licensing project manager.

## 4.6  Evaluation Findings

The reviewer will write an SER addressing each topic reviewed and explaining why the NRC staff has reasonable assurance that the radiation protection part of the application is acceptable and that the health and safety of the workers are adequately protected.  License conditions may be proposed to impose requirements where the application is deficient.  The NRC staff's SER will include the following kinds of statements and conclusions:

> The applicant has committed to an acceptable radiation protection program that includes the following:
>
> - an effective documented program to ensure that occupational radiological exposures are ALARA
>
> - an organization with adequate qualification requirements for the radiation protection personnel
>
> - approved written radiation protection procedures and RWPs for radiation protection activities
>
> - radiation protection training for all personnel who have access to restricted areas
>
> - a program to control airborne concentrations of radioactive material with engineering controls and respiratory protection
>
> - a radiation survey and monitoring program that includes requirements for controlling radiological contamination within the facility and monitoring external and internal radiation exposures
>
> - other programs to correct upsets at the facility, maintain records, and report to the NRC in accordance with 10 CFR Part 20 and 10 CFR Part 70.

The NRC staff concludes that the applicant's radiation protection program is adequate and meets the requirements of 10 CFR Part 19, 10 CFR Part 20, and 10 CFR Part 70. Conformance to the license application and license conditions will ensure safe operation.

The applicant has accurately evaluated, in the ISA Summary, those accident sequences with intermediate and high radiological consequences. The applicant has also identified controls and management measures that reduce the likelihood or consequences of accident sequences and meet the performance criteria of 10 CFR 70.61.

## 4.7 References

*U.S. Code of Federal Regulations*, Chapter I, Title 10, "Energy," Part 19, "Notices, Instructions and Reports to Workers: Inspection and Investigations."

*U.S. Code of Federal Regulations*, Chapter I, Title 10, "Energy," Part 20, "Standards for Protection Against Radiation."

*U.S. Code of Federal Regulations*, Chapter I, Title 10, "Energy," Part 70, "Domestic Licensing of Special Nuclear Material."

U.S. Nuclear Regulatory Commission, "Guide for Administrative Practice in Radiation Monitoring," Regulatory Guide 8.2, February 1973.

U.S. Nuclear Regulatory Commission, "Operating Philosophy for Maintaining Occupational Radiation Exposures as Low as Is Reasonably Achievable, Regulatory Guide 8.10, Revision 1-R, May 1977.

U.S. Nuclear Regulatory Commission, "Instructions Concerning Prenatal Radiation Exposure," Regulatory Guide 8.13, Revision 3, June 1999.

U.S. Nuclear Regulatory Commission, "Acceptable Programs for Respiratory Protection," Regulatory Guide 8.15, Revision 1, May 1977.

U.S. Nuclear Regulatory Commission, "Instructions Concerning Risks from Occupational Radiation Exposure," Regulatory Guide 8.29, Revision 1, February 1996.

U.S. Nuclear Regulatory Commission, "Minimization of Contamination and Radioactive Waste Generation: Life Cycle Planning," Regulatory Guide 4.21, June 2008.

National Council on Radiation Protection and Measurements, "Operational Radiation Safety Training," NCRP Report No. 134, 2000.

American Society for Testing and Materials, "Radiological Protection Training for Nuclear Facility Workers," ASTM E1168-95, reapproved 2008.

American National Standards Institute and the Health Physics Society, "Ionizing Radiation Safety Training for Workers," ANSI/HPS N13.36, October 30, 2001.

U.S. Nuclear Regulatory Commission, "Health Physics Surveys During Enriched Uranium-235 Processing and Fuel Fabrication," Regulatory Guide 8.24, Revision 1, October 1979.

American Conference of Governmental Industrial Hygienists, "Industrial Ventilation: A Manual of Recommended Practice for Design," ACGIH 2095, 2007.

Energy Research and Development Administration, "Nuclear Air Cleaning Handbook," ERDA 76-21, by C.A. Burchsted, A.B. Fuller, and J.E. Kahn, March 31, 1976.

American National Standards Institute, "Practices for Respiratory Protection," ANSI Z88.2-1992.

U.S. Nuclear Regulatory Commission, "Direct-Reading and Indirect-Reading Pocket Dosimeters," Regulatory Guide 8.4, February 1973.

U.S. Nuclear Regulatory Commission, "Instructions for Recording and Reporting Occupational Radiation Exposure Data," Regulatory Guide 8.7, Revision 2, November 2005.

U.S. Nuclear Regulatory Commission, "Acceptable Concepts, Models, Equations, and Assumptions for a Bioassay Program," Regulatory Guide 8.9, Revision 1, July 1993.

U.S. Nuclear Regulatory Commission, "Air Sampling in the Workplace," Regulatory Guide 8.25, Revision 1, June 1992.

U.S. Nuclear Regulatory Commission, "Monitoring Criteria and Methods To Calculate Occupational Radiation Doses," Regulatory Guide 8.34, July 1992.

U.S. Nuclear Regulatory Commission, "Air Sampling in the Workplace," NUREG-1400, September 1993.

American National Standards Institute, "Sampling and Monitoring Releases of Airborne Radioactive Substances from the Stacks and Ducts of Nuclear Facilities," ANSI N13.1-1999.

American National Standards Institute, "Radiation Protection Instrumentation Test and Calibration," ANSI N328-1978.

American National Standards Institute, "Personnel Dosimetry Performance—Criteria for Testing," ANSI N13.11-2001.

American National Standards Institute, "Radiation Detectors—Personnel Thermoluminescence Dosimetry Systems—Performance," ANSI N13.15-1985.

American National Standards Institute, "Bioassay Program for Uranium," ANSI N13.22-1995.

American National Standards Institute, "Performance Requirements for Pocket-Sized Alarm Dosimeters and Alarm Ratemeters," ANSI N13.27-1981.

American National Standards Institute, "Performance Criteria for Radiobioassay," ANSI N13.30-1996.

American National Standards Institute, "Practice for Occupational Radiation Exposure Records Systems," ANSI N13.6-1999.

U.S. Nuclear Regulatory Commission, Branch Technical Position, "License Condition for Leak-Testing Sealed Byproduct Material Sources," April 1993.

U.S. Nuclear Regulatory Commission, Branch Technical Position, "License Condition for Leak-Testing Sealed Plutonium Sources," April 1993.

U.S. Nuclear Regulatory Commission, Branch Technical Position, "License Condition for Plutonium Alpha Sources," April 1993.

U.S. Nuclear Regulatory Commission, Branch Technical Position, "License Condition for Leak-Testing Sealed Source Which Contains Alpha and/or Beta-Gamma Emitters," April 1993.

U.S. Nuclear Regulatory Commission, Branch Technical Position, "License Condition for Leak-Testing Sealed Uranium Sources," April 1993.

U.S. Nuclear Regulatory Commission, Branch Technical Position, "Guidelines for Decontamination of Facilities and Equipment Prior to Release for Unrestricted Use or Termination of Licenses for Byproduct, Source, or Special Nuclear Material," April 1993.

U.S. Nuclear Regulatory Commission, "Integrated Safety Analysis Guidance Document," NUREG-1513, May 2001.

U.S. Nuclear Regulatory Commission, "Nuclear Fuel Cycle Facility Accident Analysis Handbook," NUREG/CR-6410, March 1998.

# 5. NUCLEAR CRITICALITY SAFETY

## 5.1 Purpose of Review

The primary purpose of the review is to determine, with reasonable assurance, that the applicant has designed a facility that will provide adequate protection against criticality hazards related to the storage, handling, and processing of licensed materials, as required by Title 10 of the *Code of Federal Regulations* (10 CFR) Part 70, "Domestic Licensing of Special Nuclear Material." The facility design must adequately protect the health and safety of workers and the public during normal operations and credible accident conditions (§70.23 (a)(3))from the accidental criticality risks in the facility. It should also protect against facility conditions that could affect the safety of licensed materials and thus present an increased risk of criticality or radiation release.

Another purpose of this review is to determine, with reasonable assurance, whether the licensee's or applicant's nuclear criticality safety (NCS) program, as described in the license application and integrated safety analysis (ISA) summary, is adequate to meet the regulatory requirements in 10 CFR Part 70 and will support safe possession and use of nuclear material at the facility. The review should examine the parts of the license application and ISA Summary that describe the NCS program. The review should ensure that either the license application for a new facility or the license amendment to an existing facility meets the regulatory requirements in 10 CFR Part 70. The review should also ensure that, if applicable, the criteria specified in 10 CFR Part 70 for meeting 10 CFR 70.61, "Performance Requirements," are satisfied and that the contents of the ISA Summary required by 10 CFR 70.65, "Additional Content of Applications," meet the regulatory requirements for the NCS-related areas of the ISA Summary.

## 5.2 Responsibility for Review

Primary:     Nuclear Process Engineer (NCS Reviewer)

Supporting:  Licensing Project Manager
             Fuel Cycle Inspection staff
             Primary Reviewers for Chapters 1, 3, 8, and 11 of this Standard Review Plan
             (SRP)

## 5.3 Areas of Review

### 5.3.1 License Application

The staff should review the license application and ISA Summary, if applicable, to determine whether the application meets the 10 CFR Part 70 requirements for the NCS-related areas. The regulatory requirements for the license application review should comply with the general and additional content of an application, as required by 10 CFR 70.22, "Content of Applications," and 10 CFR 70.65, respectively. The NCS reviewer should evaluate the application or amendment to determine whether the applicant has met the requirements of 10 CFR 70.23, "Requirements for the Approval of Applications," to ensure that the applicant has proposed equipment, facilities, and procedures to protect health and minimize danger to life or property, and 10 CFR 70.64, "Requirements for New Facilities or New Processes at Existing Facilities," as applicable, to ensure that the design provides for criticality control, including adherence to the double contingency principle.

The NCS reviewer should review the ISA-related requirements in 10 CFR 70.62, "Safety Program and Integrated Safety Analysis," and 10 CFR 70.65, including the requirement for criticality monitoring and alarms. The regulation established in 10 CFR 70.62(a) requires an applicant to develop, implement, and maintain a safety program that will reasonably protect the health and safety of the public and the environment from criticality hazards associated with processing, handling, and storing licensed materials during normal operations, anticipated operational occurrences, and credible accidents. The NCS program addresses process-specific risks, as well as other criticality-related areas, such as performing calculations, making criticality evaluations, or demonstrating the subcritical margin. In addition, the NCS review should verify compliance with 10 CFR 70.61 for meeting the performance requirements and ensuring that, under normal and credible abnormal conditions, all nuclear processes are subcritical.

## 5.3.2 Nuclear Criticality Safety Program

The NCS reviewer should ensure that the applicant has committed to and implemented effective management of the NCS program in the license application and has provided enough qualified resources for an effective NCS program. The primary objective of an effective NCS program is to prevent an inadvertent nuclear criticality. Although 10 CFR Part 70 does not require a nuclear safety program directly, an applicant should provide commitments pertaining to NCS in the following areas:

- establishing and maintaining NCS safety practices and procedures

- establishing and maintaining NCS safety limits and procedures for determining operating limits

- conducting NCS evaluations to ensure that, under normal and credible abnormal conditions, all nuclear processes remain subcritical, with an approved margin of subcriticality for safety

- providing training in procedures for criticality-related possession and use of nuclear material and for response to an inadvertent nuclear criticality

- complying with NCS baseline design criteria (BDC) requirements in 10 CFR 70.64(a), if the application is for a new facility or for a new process at an existing facility

- complying with the NCS ISA Summary change process requirements of 10 CFR 70.72, "Facility Changes and Change Process"

- protecting against the occurrence of any identified accident sequence in the ISA Summary that could lead to an inadvertent nuclear criticality

- complying with the NCS performance requirements of 10 CFR 70.61

The reviewer should determine whether the applicant has identified responsibilities and authorities of individuals to implement and administer the NCS program. The reviewer should evaluate the following matters related to the applicant's organization and administration:

- the general organization and administration methods used by the applicant (see SRP Chapter 2)

- the areas of review listed in SRP Section 2.3 as they relate to NCS, including the experience, educational requirements, responsibilities, and authorities of NCS management and staff

### 5.3.3 Integrated Safety Analysis Summary

The reviewer should determine whether the applicant has committed to the facility safety program, including the process safety information, ISA, and management measures in 10 CFR 70.62, and whether the commitments demonstrate the applicant's ability to implement and maintain the NCS controls. The NCS reviewer should evaluate the following areas in the ISA Summary:

- Process descriptions are narrative descriptions of the site, facility, and processes with respect to criticality safety for normal operations. The criticality process description can include flow diagrams, major process steps, and major pieces of equipment, with emphasis on the criticality safety controls. The ISA Summary should include a reasonably simple description of each process (unit operations).

- Criticality accident sequences should include accident sequences involving licensed materials and an interpretation of the sequence of events, as described in the ISA Summary. It is assumed that all criticality accident sequences have high consequences; therefore, the applicant should include every credible event that could result in an uncontrolled criticality.

- Criticality accident consequences should be identified in the ISA Summary, including the assumption that all criticality accidents are high-consequence events and that the bases and methods the applicant used are based on using preventative controls.

- Criticality process items relied on for safety (IROFS) should include a list of items relied on for criticality safety and a description of their safety function, as described in the ISA Summary. The applicant should use enough IROFS to demonstrate that, under normal and abnormal credible conditions, all nuclear processes are subcritical.

- Criticality IROFS management measures should include management measures to ensure the reliability and availability of the IROFS described in the ISA Summary.

Review Interfaces

The criticality safety reviewer should examine information in the following other areas to ensure that it is consistent with Chapter 5 of the application:

- facility and process description applied to criticality safety, as described in Chapter 1 of this SRP

- administration and organization of the criticality safety functions, as described in Chapter 2 of this SRP

- safety program, ISA commitments, and ISA documentation applied to criticality safety under SRP Chapter 3

- emergency plan applied to criticality safety under SRP Chapter 8

- configuration management, maintenance, training and qualifications, procedures, audits and assessments, incident investigations, record management, and other quality assurance elements, as described in SRP Chapter 11

## 5.4 Acceptance Criteria

The applicant should provide NCS commitments and describe how the commitments will be met. Commitments and descriptions are expected when the acceptance criteria are relevant to the possession and use of nuclear materials and the materials to be licensed.

### 5.4.1 Regulatory Requirements

Acceptance criteria are based on meeting the relevant requirements of the following regulations:

- The general and additional contents of an application for criticality safety are in 10 CFR 70.22 and 10 CFR 70.65, respectively. General information that must be included in the license application appears in 10 CFR 70.22. Information that must be included in the ISA Summary, including the requirements for criticality monitoring and alarms, appears in 10 CFR 70.65.

- The requirements for the approval of the application are in 10 CFR 70.23.

- Requirements for new facilities or new processes at existing facilities that require a license amendment under 10 CFR 70.72 appear in 10 CFR 70.64, including the requirement to adhere to the double contingency principle.

- Requirements to maintain and establish a safety program are in 10 CFR 70.62.

- The criticality safety review should be conducted to provide reasonable assurance of compliance with the performance requirements in 10 CFR 70.61.

### 5.4.2 Regulatory Guidance

The following additional guidance may be used to supplement the review of the NCS program:

- NUREG-1513, "Integrated Safety Analysis Guidance Document," May 2001

- NUREG/CR-6410, "Nuclear Fuel Cycle Facility Accident Analysis Handbook," March 1998

### 5.4.3 Regulatory Acceptance Criteria

The reviewer should find the applicant's criticality safety program information acceptable if it provides reasonable assurance that the acceptance criteria discussed below are adequately addressed and satisfied. The applicant may elect to incorporate some or all of the requested criticality safety information in the facility and process description (SRP Section 1.1) or in the ISA Summary, rather than in this section. Either approach is acceptable, as long as the information is adequately cross-referenced.

### 5.4.3.1 License Application

The reviewer should consider that the applicant's commitment to the criticality accident alarm system (CAAS) requirements in 10 CFR 70.24, "Criticality Accident Requirements," is acceptable if the applicant or licensee has met the following acceptance criteria or has identified and justified an alternative in the application:

- The applicant describes a facility CAAS that meets the requirements of 10 CFR 70.24.

- The applicant commits to American National Standards Institute/American Nuclear Society (ANSI/ANS)-8.3-1997, "Criticality Accident Alarm System," as modified by Regulatory Guide 3.71, "Nuclear Criticality Safety Standards for Fuels and Material Facilities," issued October 2005 by the U.S. Nuclear Regulatory Commission (NRC). Regulatory Guide 3.71 lists the following exceptions to the standard:

  - At or above the 10 CFR 70.24 mass limits, the applicant should require CAAS coverage in each area where special nuclear material (SNM) is handled, stored, or used.

  - A requirement of 10 CFR 70.24 is that two detectors cover each area needing CAAS coverage.

  - A requirement of 10 CFR 70.24 is that a CAAS be capable of detecting a nuclear criticality that produces an absorbed dose in soft tissue of 20 rads (0.2 Gy) of combined neutron and gamma radiation at an unshielded distance of 2 meters from the reacting material within 1 minute.

- The applicant commits to having a CAAS that is appropriate for the facility for the type of radiation detected, intervening shielding, and the magnitude of the minimum accident of concern.

- The applicant commits to having a CAAS that is designed to remain operational during credible events, such as a seismic shock equivalent to the site-specific, design-basis earthquake or the equivalent value specified by the Uniform Building Code.

- The applicant commits to having a CAAS that is designed to remain operational during credible events, such as a fire, an explosion, a corrosive atmosphere, or other credible conditions.

- The applicant commits to having a criticality accident alarm that is clearly audible in areas that must be evacuated or provides alternative notification methods that are documented to be effective in notifying personnel that evacuation is necessary.

- The applicant commits to rendering operations safe, by shutdown and quarantine if necessary, in any area where CAAS coverage has been lost and not restored within a specified number of hours. The number of hours should be determined on a process-by-process basis, because shutting down certain processes, even to make them safe, may carry a larger risk than being without a CAAS for a short time. The applicant should commit to compensatory measures (e.g., limiting access, halting SNM movement) when the CAAS system is not functional.

- The applicant commits to the following emergency management provisions (see SRP Chapter 8):

  - The applicant commits to the requirements in ANSI/ANS-8.23-1997, "Nuclear Criticality Accident Emergency Planning and Response," as they relate to NCS.

  - The applicant either has an emergency plan or satisfies the alternative requirements in 10 CFR 70.22(h)(1)(i).

  - The applicant commits to the provision of fixed and personnel accident dosimeters in areas that require a CAAS. These dosimeters should be readily available to personnel responding to an emergency, and there should be a method for prompt onsite dosimeter readouts.

  - The applicant commits to providing emergency power for the CAAS or provides justification for the use of continuous monitoring with portable instruments.

Using the reasonable assurance of safety standard as described in the introduction to this SRP, the reviewer should determine whether the applicant has met the requirements of 10 CFR 70.61. The introduction, as well as Section 3.1 of the SRP, describing the review of the ISA and ISA Summary, includes guidance on the level of detail needed to achieve this standard. The reviewer should consider the applicant's commitments to demonstrating that all nuclear processes will be subcritical under normal and credible abnormal conditions to be acceptable if the application includes the following acceptance criteria or identifies and justifies an alternative:

- As one approach, the applicant commits to the following national standards, as they relate to these requirements: ANSI/ANS-8.7-1975, "Guide for Nuclear Criticality Safety in the Storage of Fissile Materials"; ANSI/ANS-8.9-1987, "Nuclear Criticality Safety Criteria for Steel-Pipe Intersections Containing Aqueous Solutions of Fissile Materials"; ANSI/ANS-8.10-1983, "Criteria for Nuclear Criticality Safety Controls in Operations with Shielding and Confinement"; ANSI/ANS-8.12-1987, "Nuclear Criticality Control and Safety of Plutonium-Uranium Fuel Mixtures Outside Reactors"; ANSI/ANS-8.15-1981, "Nuclear Criticality Control of Special Actinide Elements"; and ANSI/ANS-8.17-1984, "Criticality Safety Criteria for the Handling, Storage, and Transportation of LWR Fuel Outside Reactors." Alternatively, the applicant commits to base the safety limits on validated calculational methods.

- The applicant describes a program that ensures compliance with the double-contingency principle, where practicable (see Appendix 5-A for detailed guidance regarding the double-contingency principle). Processes in which there are no credible accident sequences that lead to criticality meet the double-contingency principle by definition.

This principle, as given in ANSI/ANS-8.1-1998, "Nuclear Criticality Safety in Operations with Fissionable Materials Outside Reactors," states that at least two changes in process conditions must occur before criticality is possible. If there are no process changes leading to criticality, then the principle is satisfied. Each process that has accident sequences leading to criticality should have sufficient controls in place to ensure double-contingency protection. This may be provided by either (1) control of two independent process parameters, or (2) control of a single process parameter, such that at least two independent failures would have to occur before criticality is possible. The first method is preferable, because of the inherent difficulty in preventing common-mode failure when controlling only one parameter.

The reviewer should note that the double-contingency principle requires two unlikely, independent, and concurrent changes in process conditions before criticality is possible. This does not necessarily mean that two controls are required. In some cases, it may be appropriate to credit the natural and credible course of events (e.g., unsintered powder cannot exceed a maximum density, there is no means of enriching beyond 5 weight percent uranium-235, the low historical likelihood of flooding) without establishing explicit controls. The reviewer should exercise judgment in determining whether the applicant has established sufficient means to ensure that occurrence of the contingencies is "unlikely." In addition, the term "concurrent" means that the effect of the first process change persists until a second change occurs. It does not mean that the two events must occur simultaneously. The possibility of an inadvertent criticality can be markedly reduced if failures of NCS controls are rapidly detected and the process rendered safe. If not, processes can remain vulnerable to a second failure for extended periods of time.

In a very few processes, double-contingency protection is not practicable. In those rare instances, the applicant should provide adequate justification for why such cases are acceptable. The justification should demonstrate that there is sufficiently low risk that an exception is warranted. The reviewer should note that the double-contingency principle, as stated in ANSI/ANS-8.1-1998, is a recommendation ("Process designs should, in general, incorporate sufficient factors of safety..."). The more important requirement is the one incorporated in 10 CFR 70.61(d) ("it shall be determined that the entire process will be subcritical under both normal and credible abnormal conditions"). Thus, as long as the applicant can meet the underlying requirement to be subcritical under normal and credible abnormal conditions through other means, an exception may be justifiable.

- The applicant meets the acceptance criteria in SRP Chapter 3 as they relate to the subcriticality of operations and the margin of subcriticality for safety.

The ISA and supporting ISA documentation (such as piping and instrumentation diagrams, criticality safety analyses, dose calculations, process safety information, and ISA worksheets) would be maintained on site at an existing facility. For an applicant seeking a license before completion of a facility, a full level of detail concerning hardware, procedures, and programs usually would not exist. However, at the time of the preoperational readiness review for a new facility, or a new process at an existing facility, such details must be available to demonstrate compliance with the safety program requirements of Subpart H, "Additional Requirements for Certain Licensees Authorized To Possess a Critical Mass of Special Nuclear Material," to 10 CFR Part 70.

The reviewer should consider the applicant's commitment to the BDC requirements in 10 CFR 70.64(a) acceptable if the applicant has committed to the double-contingency principle in determining NCS controls and IROFS in the design of new facilities or new processes at existing facilities that require a license amendment under 10 CFR 70.72. The applicant could also identify and justify an alternative in the application.

The applicant must meet the performance requirements in 10 CFR 70.61(b) and (c), as well as in 10 CFR 70.61(d), which include the requirement to limit the risk of inadvertent nuclear criticality by ensuring that all nuclear processes remain subcritical. The applicant's evaluation of NCS accident sequences may be performed in a manner consistent with the applicant's evaluation of non-NCS accident sequences used to meet 10 CFR 70.61(b) and (c); however, 10 CFR 70.61(d) requires the applicant to use prevention methods as the primary means to meet the performance requirements of 10 CFR 70.61(b) and (c). In addition, for new facilities and new processes at existing facilities, 10 CFR 70.64(a)(9) requires compliance with the double-contingency principle. This requires considerations in addition to those necessary to meet 10 CFR 70.61 for the noncriticality hazards.

The reviewer should consider the applicant's commitment to the requirements in 10 CFR 70.65(b) acceptable if it has met the following acceptance criteria or has identified and justified an alternative in the application:

- The applicant meets the acceptance criteria in SRP Section 3, as they relate to the identification of NCS accident sequences, consequences of NCS accident sequences, likelihoods of NCS accident sequences, and descriptions of IROFS for NCS accident sequences.

- The applicant should consider the upsets listed in Appendix A to ANSI/ANS-8.1-1983 in identifying NCS accident sequences.

The applicant may use the guidance in ANSI/ANS-8.10-1983, as modified by Regulatory Guide 3.71, in determining the consequences of criticality accident sequences. In general, such events should be considered "high-consequence" events unless controls are in place to provide shielding or other isolation between the source of radiation and facility personnel. Consideration of events as other than those of high consequence should be justified in the ISA Summary. The reviewer should note that the requirements of 10 CFR 70.61(d) are still applicable (i.e., criticality is to be prevented).

The application should also address the BDC for new facilities or new processes at existing facilities that require a license amendment under 10 CFR 70.72. These baseline criteria must (§70.64 (a)(9))be applied to the design of new processes but do not require retrofits to existing facilities or existing processes; however, all facilities and processes must comply with the performance requirements in 10 CFR 70.61. Section 2.4 of NUREG-1601, "Chemical Process Safety at Fuel Cycle Facilities," dated September 3, 1997, contains a list of items that should be considered in an adequate facility design. For new facilities and new processes in existing facilities, the design should provide for adequate protection against criticality accidents.

The application should do the following:

- The applicant describes how it performed the ISA for the new process and how the ISA satisfies the principles of the BDC and the performance requirements in 10 CFR 70.61. The applicant also explains how it applies defense in depth to higher risk accident sequences. Acceptable defense-in-depth principles for the criticality safety design are those that support a hierarchy of controls: prevention, mitigation, and operator intervention, in order of preference.

- The applicant describes proposed facility-specific or process-specific relaxations or additions to BDC, along with justifications for relaxations.

- The ISA Summary describes how the criticality safety BDC were applied in establishing the design principles, features, and control systems of the new process.

### 5.4.3.2 Nuclear Criticality Safety Program

The reviewer should consider the applicant's management of the NCS program acceptable if the applicant has met the following acceptance criteria or has identified and justified an alternative in the application:

- The applicant describes and commits to implementing and maintaining an NCS program to meet the regulatory requirements of 10 CFR Part 70.

- The application states the NCS program objectives, which should include those listed in this chapter.

- The application outlines an NCS program structure that is consistent with current industry practices (e.g., ANSI/ANS-8.1-1998 and ANSI/ANS-8.19-1996, "Administrative Practices for Nuclear Criticality Safety") and current industry practice that defines the responsibilities and authorities of key program personnel.

- The applicant commits to using the NCS program to establish and maintain NCS safety limits and NCS operating limits for fissile material use and possession and commits to maintaining management measures to ensure the availability and reliability of the controls.

- The applicant commits to preparing NCS postings, to NCS training, and to NCS procedure training.

- The applicant commits to evaluating modifications to the facility or safety program for their impact on criticality safety.

The organization and administration part of the application (see SRP Chapter 2) contains information related to NCS organization and administration acceptance criteria. The reviewer should find the applicant's NCS organization and administration acceptable if the applicant has met the following acceptance criteria or has identified and justified an alternative in the application:

- The applicant meets the acceptance criteria in SRP Section 2.4 as they relate to NCS, including organizational positions, functional responsibilities, experience, and qualifications of personnel responsible for NCS.

- The applicant meets the intent of ANSI/ANS-8.1 and ANSI/ANS-8.19 (see Regulatory Guide 3.71), as they relate to organization and administration.

- The NCS organization should be independent of operations to the extent practical.

- The applicant commits to providing distinctive NCS postings in areas, operations, work stations, and storage locations relying on administrative controls for NCS.

- The applicant commits to requiring its personnel to perform activities in accordance with written, approved procedures when the activity may affect NCS. Unless a specific procedure deals with the situation, personnel shall take no action until the NCS staff has evaluated the situation and provided recovery procedures.

- The applicant commits to requiring its personnel to report defective NCS conditions to the NCS program management.

- The applicant describes organizational positions, experience of personnel, qualifications of personnel, and functional responsibilities.

- The applicant commits to designating an NCS program director who will be responsible for implementing the NCS program.

Information related to acceptance criteria for the NCS safety program may appear in the ISA or management measures part of the application. The applicant's NCS management measures (required by 10 CFR 70.62) should be considered acceptable if the applicant has met the following acceptance criteria or has identified and justified an alternative in the application:

- training (see SRP Chapter 11)

  - The applicant meets the intent of ANSI/ANS-8.19 and ANSI/ANS-8.20, "Nuclear Criticality Safety Training," as they relate to training.

  - The applicant commits to training all personnel to recognize the CAAS signal and to evacuate promptly to a safe area.

  - The applicant commits to providing instruction and training regarding the policy in the SRP guidance for NCS organization procedures (see SRP Chapter 11).

- procedures (see SRP Chapter 11)

  - The applicant commits to ANSI/ANS-8.19-1996 as it relates to procedures.

- audits and assessments (see SRP Chapter11)

  - The applicant commits to ANSI/ANS-8.19-1996 as it relates to audits and assessments.

  - The applicant commits to conducting and documenting walkthroughs (i.e., observation of operations to ensure compliance with criticality limits) of all operating SNM process areas, such that all operating SNM process areas will be reviewed at some specified frequency. The reviewer should consider the complexity of the process, the degree of process monitoring, and the degree of reliance on administrative controls in assessing the acceptability of the specified frequency. Identified weaknesses should be referred to those responsible for facility corrective actions and should be promptly and effectively resolved. A graded approach may be used to justify an alternative NCS walkthrough schedule.

– The applicant commits to conducting and documenting periodic NCS audits (such that all NCS aspects of management measures (see SRP Chapter 11) will be audited at least every 2 years). A graded approach may be used to justify an alternative NCS audit schedule.

The reviewer should consider the applicant's NCS technical practices acceptable if the applicant has met the following acceptance criteria or has identified and justified an alternative in the application:

• NCS evaluations will be performed using industry-accepted and peer-reviewed methods.

• NCS limits on controlled parameters will be established to ensure that all nuclear processes are subcritical, including an adequate margin of subcriticality for safety.

• Methods used to develop NCS limits will be validated to ensure that they are used within acceptable ranges and that the applicant used both appropriate assumptions and acceptable computer codes.

• The applicant commits to demonstrating (1) the adequacy of the margin of subcriticality for safety by ensuring that the margin is large compared to the uncertainty in the calculated value of Keff (effective multiplication factor), (2) that the calculation of Keff is based on a set of variables within the method's validated area of applicability, and (3) that trends in the bias support the extension of the methodology to areas outside the area or areas of applicability.

The margin of subcriticality for safety is an allowance for any unknown uncertainties that have not been accounted for in validation and a measure of the degree of confidence that systems calculated to be subcritical are actually subcritical. The margin is used to define an upper subcritical limit, as follows:

k-subcritical = 1.0 – bias – bias uncertainty – margin of subcriticality for safety

The reviewer must use judgment in assessing whether the margin of subcriticality for safety is sufficient to provide reasonable assurance of subcriticality (in accordance with 10 CFR 70.61(d)). The reviewer should consider the following factors, as applicable, in making this judgment, as well as any other available information that provides the needed confidence:

– conservatism in the calculations, beyond that needed to accommodate uncertainties in the modeled parameters (e.g., geometric tolerances)

– confidence in subcriticality generated by the applicant's validation process, including the following:

- similarity between the benchmark experiments and calculations to be performed

- sufficiency of the benchmark data (both quality and quantity)

- rigor of the validation methodology (e.g., trending, statistical testing)

- conservatism in the statistical parameters (e.g., 95/95 lower tolerance limit)

– sensitivity of the system to changes in modeled parameters (and therefore to errors)

– corroborating evidence of subcriticality from other sources (e.g., knowledge of neutron physics for well-characterized systems, such as finished fuel)

– risk considerations, including the likelihood of actually attaining an abnormal condition

In general, a margin of subcriticality for safety of 0.05 has been found acceptable for typical nuclear processes involving low-enriched uranium, without a detailed justification. The use of increasingly smaller margins should require increasingly more rigorous justification, and the reviewer should evaluate other physical systems on a case-by-case basis.

- The applicant includes a summary description of a documented, reviewed, and approved validation report (by NCS function and management) for each methodology that will be used to perform an NCS analysis (e.g., experimental data, reference books, hand calculations, deterministic computer codes, probabilistic computer codes). The summary description of a reference manual or validation report should include the following:

– a summary of the theory of the methodology that is sufficiently detailed and clear to be understood, including the method used to select the benchmark experiments, determine the bias and uncertainty in the bias, and determine the upper subcritical limit

– a summary of the physical systems and area(s) of applicability covered by the validation report, noting that it is not necessary to include the full range of numerical parameters that defines the area of applicability. since a general description (e.g., low-enriched homogeneous uranyl fluoride solutions, low-enriched fuel pellets, and rods containing gadolinia) is sufficient

– a description of the methods used to justify applying the methodology outside the area or areas of applicability

– a summary of the plant-specific benchmark experiments used to validate the methodology, noting that it is not necessary to include all benchmark experiments used, since a brief description of the individual benchmark data sets will suffice

–   a description of the margin of subcriticality for safety and its justification

–   a description of the controlled software and hardware used

–   a description of the verification process used, including verification upon changes to the calculational system and upon some specified period

- The applicant's validation methodology, as described above, should be found acceptable if either (1) the applicant commits to following ANSI/ANS-8.24-2006, "Validation of Neutron Transport Methods for Nuclear Criticality Safety Calculations," as endorsed by Regulatory Guide 3.71, or (2) the methodology follows current industry practices in terms of selecting the benchmark experiments, assessing their applicability, determining the area(s) of applicability, extending the area(s) of applicability beyond the range of benchmark data, and statistically analyzing the data. This requires that the NCS reviewer remain aware of current practices in the area of criticality code validation.

The reviewer may examine the applicant's validation report to ensure that the methodology is sufficiently rigorous and is being applied in a manner consistent with its assumptions (e.g., normal distribution of benchmarks).

–   The applicant commits to incorporating each validation report into the facility configuration management program.

–   The applicant commits to performing NCS analyses in accordance with documented and approved procedures, which incorporate the following principles:

  - NCS safety limits and NCS operating limits will be established, assuming optimum credible conditions (i.e., the most reactive conditions physically possible or limited by written commitments to regulatory agencies) unless specified controls are implemented to control the limit to a certain range of values.

  - NCS safety limits, NCS operating limits, and limits on NCS-controlled parameters will be derived from the NCS analyses.

  - NCS operating limits will be derived from NCS safety limits by considering the uncertainty and variability in operating parameters to ensure that processes will remain subcritical under both normal and credible abnormal conditions.

  - The margin of subcriticality for safety for a process should be large, relative to the uncertainty in the calculated value of Keff.

Controlled parameters available for NCS control include the following: mass, geometry, density, enrichment, reflection, moderation, concentration, interaction, neutron absorption, and volume. The reviewer should consider the applicant's commitment to NCS technical practices acceptable if the applicant has met the following acceptance criteria or has identified and justified an alternative in the application:

- The applicant's use of a single NCS control to maintain the values of two or more controlled parameters constitutes only one component necessary to meet double-contingency protection.

- The applicant commits to the preferred use of passive-engineered controls to ensure NCS. In general, the applicant should commit to the following order of preference for NCS controls: (1) passive engineered, (2) active engineered, (3) enhanced administrative, and (4) simple administrative. When using other than a passive-engineered control, the applicant should justify the choice of the type and manner.

- When they are relevant, the applicant should consider heterogeneous effects. Heterogeneous effects are particularly relevant for low-enriched uranium processes, where, all other parameters being equal, heterogeneous systems are more reactive than homogeneous systems.

The use of mass as a controlled parameter should be considered acceptable in the following circumstances:

- When mass limits are derived for a material that is assumed to have a given weight percent of SNM, determinations of mass are based on either (1) weighing the material and assuming that the entire mass is SNM or (2) conducting physical measurements to establish the actual weight percent of SNM in the material.

- When fixed geometric devices are used to limit the mass of SNM, a conservative process density is assumed in calculating the resulting mass.

- When the mass is measured, instrumentation subject to facility management measures is used.

The use of geometry as a controlled parameter should be considered acceptable if the following applies:

- Before beginning operations, all dimensions and nuclear properties that use geometry control are verified. The facility configuration management program should be used to maintain these dimensions and nuclear properties.

The use of density as a controlled parameter should be considered acceptable in the following circumstances:

- When process variables can affect the density, the ISA Summary shows the process variables to be controlled by IROFS.

- Density is measured by the use of instrumentation subject to facility management measures.

The use of enrichment as a controlled parameter should be considered acceptable if the following apply:

- Either a method of segregating enrichments is used to ensure that differing enrichments will not be interchanged or the most limiting enrichment is applied to all material.

- Measurements of enrichment are obtained by using instrumentation subject to facility management measures.

The use of reflection as a controlled parameter should be considered acceptable in the following circumstances:

- In the evaluation of an individual unit, the wall thickness of the unit and all reflecting adjacent materials of the unit are considered. The materials adjacent to the unit should be farther than 30 centimeters (12 inches).

- After all fixed reflectors are accounted for, the controls to prevent the presence of any transient reflectors (e.g., personnel) are identified as IROFS in the ISA Summary.

The use of moderation as a controlled parameter should be considered acceptable if the following apply:

- When using moderation, the applicant commits to ANSI/ANS-8.22-1997, "Nuclear Criticality Safety Based on Limiting and Controlling Moderators."

- When process variables can affect the moderation, the ISA Summary shows the process variables to be controlled by IROFS.

- Moderation is measured by using instrumentation subject to facility management measures.

- The design of physical structures prevents the ingress of moderators.

- When moderation needs to be sampled, dual independent sampling methods are used.

- Firefighting procedures for use in a moderation-controlled area evaluate the use of moderator material.

- After evaluation of all credible sources of moderation for the potential for intrusion into a moderation-controlled area, the ingress of moderation is prevented or controlled.

The use of concentration as a controlled parameter should be considered acceptable in the following circumstances:

- When process variables can affect the concentration, the ISA Summary shows the process variables to be controlled by IROFS.

- Concentrations of SNM in a process are limited unless the process is analyzed to be safe at any credible concentration.

- When using a tank containing concentration-controlled solution, the tank is normally closed and locked to prevent unauthorized access.

- When concentration needs to be sampled, dual independent sampling methods are used.

- After identification of possible precipitating agents, precautions are taken to ensure that such agents will not be inadvertently introduced.

The use of interaction as a controlled parameter should be considered acceptable if the following applies:

- To maintain a physical separation between units, engineered controls are used to ensure a minimum spacing. If engineered controls are not feasible, augmented administrative controls are used.

- The structural integrity of the spacers or racks should be sufficient for normal and credible abnormal conditions.

The use of neutron absorption as a controlled parameter should be considered acceptable in the following circumstances:

- When using borosilicate-glass raschig rings, the applicant commits to ANSI/ANS-8.5-1996, "Use of Borosilicate-Glass Raschig Rings as a Neutron Absorber in Solutions of Fissile Material."

- When using fixed neutron absorbers, the applicant commits to ANSI/ANS-8.21-1995, "Use of Fixed Neutron Absorbers in Nuclear Facilities Outside Reactors."

- In the evaluation of absorber effectiveness, neutron spectra are considered (e.g., cadmium is an effective absorber for thermal neutrons but ineffective for fast neutrons).

The use of volume as a controlled parameter should be considered acceptable if the following apply:

- Fixed geometry is used to restrict the volume of SNM.

- When the volume is measured, the instrumentation used is subject to facility management measures.

The reviewer should consider the applicant's description of additional commitments for the NCS program acceptable if the applicant has met the following acceptance criteria or has identified and justified an alternative in the application:

- The applicant commits to using the NCS program to promptly detect any NCS deficiencies by means of operational inspections, audits, or investigations and to refer to those responsible for the facility's corrective actions any unacceptable performance deficiencies in IROFS, NCS function, or management measures, so as to prevent recurrence.

- The applicant commits to supporting the facility change mechanism process by performing NCS evaluations to determine changes to processes, operating procedures, criticality controls, IROFS, and management measures.

- The applicant commits to retaining records of NCS deficiencies and to documenting any corrective actions taken.

- The reviewer should consider the applicant's description of measures to implement the facility change process requirements in 10 CFR 70.72 acceptable if the applicant has met the following acceptance criteria or has identified and justified an alternative in the application:

    - The applicant describes a change control process that is sufficient to ensure that the safety basis of the facility will be maintained during the lifetime of the facility. The change process should be documented in written procedures and should ensure that all changes to SNM processes are evaluated to determine the effect of the change on the safety basis of the process, including the effect on bounding process assumptions, the reliability and availability of NCS controls, and the NCS of connected processes. The change control process should include procedures for the review and approval of facility changes by the NCS function to determine the potential effects on NCS.

    - The change control process should be connected to the facility's configuration management system to ensure that changes to the NCS basis are incorporated into procedures, evaluations, postings, drawings, other safety-basis documentation, and the ISA Summary.

- The applicant's description of measures to implement the reporting requirements in Appendix A, "Reportable Safety Events," to 10 CFR Part 70 for criticality safety-related commitments should be considered acceptable if the commitments are consistent with the overall Appendix A program commitments and the applicant has met the following acceptance criteria or has identified and justified an alternative in the application:

    - The applicant has a program for evaluating the criticality significance of NCS events and an apparatus in place for making the required notification to the NRC Operations Center. Qualified individuals should make the determination of significance of NCS events. The determination of loss or degradation of double-contingency protection should be made against the license and Appendix A to 10 CFR Part 70.

    - The applicant incorporates the reporting criteria of Appendix A to 10 CFR Part 70 and the report content requirements of 10 CFR 70.50, "Reporting Requirements," into the facility emergency procedures.

    - The applicant commits to issuing the necessary report, based on whether the criticality controls and IROFS credited were lost (i.e., they were unreliable or unavailable to perform their intended safety functions), irrespective of whether the safety limits of the associated parameters were actually exceeded.

    –    If the licensee cannot ascertain within 1 hour whether the criteria of
10 CFR Part 70, Appendix A, paragraph (a) or (b) apply, the applicant commits to
treating the event as a 1-hour reportable event.

The applicant may use standards as a means to meet regulatory requirements. Regulatory Guide 3.71 endorses the ANSI/ANS-8 national standards, with some exceptions. The NRC endorsement of these standards means that they provide procedures and methodology generally acceptable to the NRC staff for the prevention and mitigation of nuclear criticality accidents. However, application of a standard is not a substitute for detailed NCS analyses for specific operations.

If the applicant intends to conduct activities to which an NRC-endorsed standard applies, the applicant should meet the intent of the standard by satisfying the following acceptance criteria:

- The license application contains a commitment to follow the requirements (i.e., "shall" statements) of the standard, subject to any exceptions taken by the NRC. The application clearly specifies the version of the standard and the specific provisions to which the applicant is committing.

- If there are requirements in a standard that the applicant does not commit to, the applicant provides sufficient information for the staff to determine if the requirements are not relevant to the applicant's activities or the license application contains other commitments that are equivalent.

If the licensee commits to a standard that the NRC has not endorsed, is not the most current version endorsed by the NRC, or is an unendorsed version of a previously endorsed standard, the license application should include justification for this commitment.

Regulatory Guide 3.71 endorses, in part or in full, the following ANSI/ANS-8 national standards:

- ANSI/ANS-8.1-1998

- ANSI/ANS-8.3-1997

- ANSI/ANS-8.5-1996

- ANSI/ANS-8.6, "Safety in Conducting Subcritical Neutron-Multiplication Measurements In Situ," 1983 (reaffirmed in 1995)

- ANSI/ANS-8.7-1975 (reaffirmed in 1987)

- ANSI/ANS-8.9-1987 (reaffirmed in 1995)

- ANSI/ANS-8.10-1983 (reaffirmed in 1988)

- ANSI/ANS-8.12-1987 (reaffirmed in 1993)

- ANSI/ANS-8.15-1981 (reaffirmed in 1995)

- ANSI/ANS-8.17-1984 (reaffirmed in 1997)

- ANSI/ANS-8.19-1996

- ANSI/ANS-8.20-1991

- ANSI/ANS-8.21-1995

- ANSI/ANS-8.22-1997

- ANSI/ANS-8.23-1997

- ANSI/ANS-8.24-2007

- ANSI/ANS-8.26, "Criticality Safety Engineer Training and Qualification Program," 2007

### 5.4.3.3 Integrated Safety Analysis Summary

The reviewer should find the applicant's criticality safety information acceptable if it provides reasonable assurance that the acceptance criteria presented below are adequately addressed and satisfied. The applicant may elect to incorporate some or all of the requested process information in the facility and process description (SRP Section 1.1) or the ISA Summary, rather than in this section. Either approach is acceptable, as long as the information is adequately cross-referenced.

The regulation in 10 CFR 70.65(b)(3) requires, in the ISA Summary, a description of each process in the facility. The applicant's descriptions of the criticality processes are acceptable if they meet the conditions described below.

5.4.3.3.1 Criticality Process Description

Process descriptions are sufficiently detailed to allow an understanding of the criticality to allow development of potential accident sequences.

Process descriptions are sufficiently detailed to allow an understanding of the theory of operation.

5.4.3.3.2 Criticality Accident Sequences

The use of accident sequences to demonstrate compliance is acceptable in the following circumstances:

- The applicant provides a general description of the accident sequences identified in the ISA process for criticality hazards.

- The ISA Summary describes the hazards identified in the ISA. Each accident sequence identified by the applicant in the ISA should include a criticality hazard evaluation of potential interactions and key assumptions, vessels, process equipment, and facility personnel. The hazard evaluation should use appropriate accepted methods.

- The applicant provides reasonable assurance that measures to mitigate the consequences of accident sequences identified in the ISA Summary are consistent with actions described in SRP Chapter 8. (Note that some facilities are not required to have an emergency plan.) Preventive controls and measures should be the primary means of protection against nuclear criticality accidents (§70.61(d)).

- All the credible criticality accident sequences should be assumed to have high consequences.

### 5.4.3.3.3 Criticality Items Relied on for Safety

The regulation in 10 CFR 70.65(b)(6) requires a list briefly describing all IROFS in sufficient detail to understand their functions in relation to the performance requirements.

The applicant provides, in the ISA Summary, a list of criticality safety controls (i.e., IROFS) suitable to prevent criticality accidents. This list should also briefly describe the IROFS, in sufficient detail to permit an understanding of their safety functions. The applicant should demonstrate that the likelihood of each credible high-consequence event will be highly unlikely.

If the applicant takes a graded approach to safety, in accordance with 10 CFR 70.62(a), the reviewer should establish that the grading of IROFS is appropriate and sufficient to protect against criticality risks. For example, the applicant should consider reliance on passive controls of active systems and defense in depth, in accordance with 10 CFR 70.64(b). To reduce common-mode failures, the applicant should favor design features that use independent sources of motive force.

### 5.4.3.3.4 Management Measures

The applicant should review management measures to ensure the availability and reliability of IROFS when they are required to perform their safety functions. Management measures may be graded commensurate with risk. The regulation in 10 CFR 70.65(b)(4) requires information that demonstrates the licensee's compliance with the performance requirements, including a description of the management measures.

The application should meet the following criteria:

- The application should describe the engineering approach, basis, or schemes employed for maintaining safety in normal operations.

- The ISA Summary must identify the administrative and engineered controls to prevent a criticality hazard. The applicant should also explain how any safety grading of IROFS and management measures has been made and how such grading is commensurate with the reduction in risk that the IROFS are designed to achieve.

- The application should demonstrate the management measures proposed to ensure that IROFS are available and reliable by briefly describing the following:

    - procedures to ensure the reliable operation of engineered controls
      (e.g., inspection and testing procedures and frequencies, calibration programs,

functional tests, corrective and preventive maintenance programs, criteria for acceptable test results)

—   procedures to ensure that administrative controls will be correctly implemented, when required (e.g., employee training and qualification in operating procedures, refresher training, safe work practices, development of standard operating procedures, training program evaluation)

—   the configuration management, maintenance, training and qualifications, procedures, audits and assessments, incident investigations, records management, and other quality assurance elements used by the applicant (see SRP Sections 11.3.1 through 11.3.8)

—   management provisions for the following:

- training and qualifications of NCS management and staff

- auditing, assessing, and upgrading the NCS program

- maintaining current NCS safety-basis documentation

- installing and maintaining a CAAS to detect and annunciate an inadvertent nuclear criticality

- referring NCS deficiencies to the corrective action program

- retaining records of the NCS program, including independent reviews, audits, and documentation of corrective actions taken

### 5.4.3.3.5 Requirements for New Facilities or New Processes at Existing Facilities

The application should address the BDC for new facilities or new processes at existing facilities that require a license amendment under 10 CFR 70.72. The baseline criteria must be applied to the design of new processes but do not require retrofits to existing facilities or existing processes (§70.64(a)); however, all facilities and processes must comply with the performance requirements in 10 CFR 70.61.

The applicant should state clearly how the design of the new facility or process provides for criticality control, as required in 10 CFR 70.64(a)(9). The discussion should identify how the following were considered in the design:

- subcriticality under normal and abnormal conditions
- CAAS
- implementation of double contingency

The licensee could indicate its preference in the selection of controls, such as the following:

- engineered over administrative controls

- favorable geometry design
- two-parameter control

## 5.5 Review Procedures

The reviewer should use the regulatory guidance of this chapter, references in this chapter, and the applicant's reports to the NRC (e.g., NRC Bulletin 91-01, "Reporting Loss of Criticality Safety Controls"; 10 CFR 70.50; and 10 CFR 70.74, "Additional Reporting Requirements").

### 5.5.1 Acceptance Review

The primary reviewer should review the applicant's NCS information for completeness with respect to the requirements in 10 CFR 70.22, 10 CFR 70.24, 10 CFR 70.61, 10 CFR 70.62, 10 CFR 70.64, and 10 CFR 70.65, and the acceptance criteria in Section 5.4. If deficiencies are identified, then either the reviewer should ask the applicant to submit additional material before the start of the safety evaluation or the application should be denied.

### 5.5.2 Safety Evaluation

After the application has been accepted, the primary reviewer should conduct a complete review of the application and determine if it meets the requirements for approval specified in Section 5.4. The primary reviewer should consult with the supporting reviewers, as appropriate, to identify and resolve any issues of concern related to the licensing review. The primary reviewer should also coordinate with other primary reviewers of SRP Chapters 2, 3, 8, and 11 to confirm that the application meets all acceptance criteria pertinent to NCS. The reviewer should also coordinate with other primary reviewers in radiation protection, chemical safety, and fire protection, as well as other disciplines as appropriate (e.g., seismic), to ensure appropriate consideration of any cross-cutting issues.

#### 5.5.2.1 License Application

The primary reviewer should review the applicant's NCS information in the license application for completeness with respect to the requirements in 10 CFR 70.22, 10 CFR 70.23, 10 CFR 70.24, 10 CFR 70.61, 10 CFR 70.62, 10 CFR 70.64, and 10 CFR 70.65, and the acceptance criteria in Section 5.4.

During the license application review the reviewer should identify and note any items or issues that should be inspected during an operational readiness review, if such a review will be performed. These items could include confirming that the commitments made in the license application are implemented though procedures and training.

If, during the review, the primary reviewer determines a need for additional information, the reviewer coordinates a request for additional information with the licensing project manager. The reviewer should ascertain that the criticality safety approach is consistent with other sections of the application, including those addressed by SRP Chapters 2, 3, 4, 6, 8, and 11.

For an existing facility, the reviewer may consult NRC inspectors to identify and resolve any issues related to the licensing review. These interactions should be coordinated through the licensing project manager.

The primary reviewer will prepare safety evaluation report (SER) input for the licensing project manager in support of the licensing action.

### 5.5.2.2 Nuclear Criticality Safety Program

The reviewer should review all aspects of the applicant's NCS program, including management, organization, and technical practices. The reviewer should identify and note any items or issues relating to the NCS program and commitments that should be inspected during an operational readiness review, if such a review will be performed. These items could include confirming that the commitments made in the license application are implemented through procedures and training.

If, during the review, the primary reviewer determines a need for additional information regarding the NCS program, the reviewer coordinates a request for additional information with the licensing project manager.

For an existing facility, the reviewer may consult NRC inspectors to identify and resolve any issues related to the NCS program commitments. These interactions should be coordinated through the licensing project manager.

### 5.5.2.3 Integrated Safety Analysis Summary

The results of the ISA support the overall safety basis for the criticality safety evaluation. The reviewer should assess the criticality safety risks identified in the ISA Summary and ensure that the level of safety is reflected in the design and the operational plans for the facility. The reviewer should establish that the applicant's facility design, operations, and IROFS for criticality safety provide reasonable assurance that they will function as intended, be reliable and available to perform their safety function, and provide for the safe possession and use of licensed material at the facility.

## 5.6 Evaluation Findings

SRP Chapter 3 contains the evaluation findings for the ISA Summary requirements for 10 CFR 70.65.

If the staff's review verifies that the safety program description presents sufficient information to satisfy the acceptance criteria in SRP Section 5.4, the staff may document its review as follows:

The staff has reviewed the Nuclear Criticality Safety (NCS) program and requirements for criticality safety for [name of facility] according to SRP Chapter 5. The staff has reasonable assurance of the following:

- The applicant will have in place a staff of managers, supervisors, engineers, process operators, and other support personnel who are qualified to develop, implement, and maintain the NCS program in accordance with the facility organization and administration and management measures.

- The applicant's conduct of operations will be based on NCS technical practices, which will ensure that the fissile material will be possessed, stored, and used safely, according to the requirements in 10 CFR Part 70.

- The applicant will develop, implement, and maintain a criticality accident alarm system in accordance with both the requirements in 10 CFR 70.24 and the facility emergency management program.

- The applicant will have in place an NCS program that meets the performance requirements in 10 CFR 70.61(b), the subcriticality requirement in 10 CFR 70.61(d), and the baseline design criteria requirements in 10 CFR 70.64(a).

- Based on this review, the staff concludes that the applicant's NCS program meets the requirements of 10 CFR Part 70 and provides reasonable assurance of the protection of public health and safety, including that of workers, and the environment.

## 5.7 References

*U.S. Code of Federal Regulations*, Chapter I, Title 10, "Energy," Part 70, "Domestic Licensing of Special Nuclear Material."

U.S. Nuclear Regulatory Commission, "Integrated Safety Analysis Guidance Document," NUREG-1513, May 2001.

U.S. Nuclear Regulatory Commission, "Nuclear Fuel Cycle Facility Accident Analysis Handbook," NUREG/CR-6410, March 1998.

American National Standards Institute/American Nuclear Society, "Criticality Accident Alarm System," ANSI/ANS-8.3-1997.

U.S. Nuclear Regulatory Commission, "Nuclear Criticality Safety Standards for Fuels and Material Facilities," Regulatory Guide 3.71, October 2005.

American National Standards Institute/American Nuclear Society, "Nuclear Criticality Accident Emergency Planning and Response," ANSI/ANS-8.23-1997.

American National Standards Institute/American Nuclear Society, "Guide for Nuclear Criticality Safety in the Storage of Fissile Materials," ANSI/ANS-8.7-975.

American National Standards Institute/American Nuclear Society, "Nuclear Criticality Safety Guide for Pipe Intersections Containing Aqueous Solutions of Enriched Uranyl Nitrate," ANSI/ANS-8.9-1987.

American National Standards Institute/American Nuclear Society, "Criteria for Nuclear Criticality Safety Controls in Operations with Shielding and Confinement," ANSI/ANS-8.10-1983.

American National Standards Institute/American Nuclear Society, "Nuclear Criticality Control and Safety of Plutonium-Uranium Fuel Mixtures Outside Reactors," ANSI/ANS-8.12-1987.

American National Standards Institute/American Nuclear Society, "Nuclear Criticality Control of Special Actinide Elements," ANSI/ANS-8.15-1981.

American National Standards Institute/American Nuclear Society, "Criticality Safety Criteria for the Handling, Storage, and Transportation of LWR Fuel Outside Reactors," ANSI/ANS-8.17-1984.

American National Standards Institute/American Nuclear Society, "Nuclear Criticality Safety in Operations with Fissionable Materials Outside Reactors," ANSI/ANS-8.1-1998.

U.S Nuclear Regulatory Commission, "Chemical Process Safety at Fuel Cycle Facilities," NUREG-1601, September 3, 1997.

American National Standards Institute/American Nuclear Society, "Administrative Practices for Nuclear Criticality Safety," ANSI/ANS-8.19.

H.K. Clark, "Maximum Safe Limits for Slightly Enriched Uranium and Uranium Oxide," DP-1014, Du Pont de Nemours and Co., Aiken, SC, 1966.

R.A. Knief, "Nuclear Criticality Safety—Theory and Practice," American Nuclear Society, La Grange Park, IL, 1985.

H.C. Paxton and N.L. Pruvost, "Critical Dimensions of Systems Containing $^{235}$U, $^{239}$Pu, and $^{233}$U," LA-10860-MS, Los Alamos National Laboratory, Los Alamos, NM, 1987.

N.L. Pruvost and H.C. Paxton, "Nuclear Criticality Safety Guide," LA-12808/UC-714, Los Alamos National Laboratory, Los Alamos, NM, 1996.

W.R. Stratton (D.R. Smith Revisor), "A Review of Criticality Accidents," DOE/NCT-04, U.S. Department of Energy, March 1989.

U.S. Department of Energy, "Facility Safety," DOE Order 420.1 (Change 2), October 24, 1996.

# APPENDIX A

## NUCLEAR CRITICALITY SAFETY PERFORMANCE REQUIREMENTS
## AND DOUBLE CONTINGENCY PRINCIPLE

Title 10 of the Code of Federal Regulations (10 CFR) Part 70, Subpart H contains three separate requirements to ensure nuclear criticality safety. One requirement, 10 CFR 70.64(a)(9), requires that the design of new facilities and processes provide for criticality control including adherence to the double contingency principle. A second requirement, 10 CFR 70.61(b), requires that high consequence events (which typically will include criticality accidents) be highly unlikely. A third requirement, 10 CFR 70.61(d), requires that nuclear criticality accidents be limited by assuring that under normal and abnormal conditions all nuclear processes are subcritical, including use of an approved margin of subcriticality, and also requires that the primary means of criticality protection be prevention.

The purpose of this appendix is to clarify the relationship between these three requirements.

### Discussion

There are three separate requirements in 10 CFR Part 70 for ensuring nuclear criticality safety. The first requirement of 10 CFR 70.64(a)(9) is more prescriptive and deterministic than the performance requirements of 10 CFR 70.61. 10 CFR 70.64 establishes baseline design criteria for new facilities and processes, similar to general design criteria in 10 CFR Part 50. One of these baseline design criteria applies directly to criticality safety. Specifically, 10 CFR 70.64(a)(9) requires that the design "provide for criticality control including adherence to the double contingency principle." Section 70.64(b) further specifies that new facilities or processes must incorporate defense-in-depth practices, which is defined as a "design philosophy, applied from the outset and through completion of the design, that is based on providing successive levels of protection such that health and safety will not be wholly dependent upon any single element of the design, construction, maintenance, or operation of the facility." Section 70.64(b)(1) specifically mentions preference for the selection of engineered controls over administrative controls to increase overall system reliability.

Another more risk-informed and performance-based requirement is contained in 10 CFR 70.61. In short, this regulation stipulates that credible high consequence events shall be made "highly unlikely" or be mitigated (10 CFR 70.61(b)) and that intermediate consequences shall be made "unlikely" or be mitigated (10 CFR 70.61(c)). High and intermediate consequence thresholds for workers and members of the public are established for both chemical and radiological events. Under this risk-informed and performance-based regulation a criticality accident would typically be considered a high consequence event to the worker since the worker could receive a dose in excess of 100 rem TEDE (total effective dose equivalent).

In addition, there is a separate provision within 10 CFR 70.61 that specifically addresses criticality safety. Section 70.61(d) states that, in addition to meeting the requirements above for high and intermediate consequence events, the "risk of nuclear criticality accidents must be limited by assuring that under normal and credible abnormal conditions, all nuclear processes are subcritical, including use of an approved margin of subcriticality for safety. Preventive controls and measures must be the primary means of protection against nuclear criticality accidents." The purpose of this is to preclude a situation where nuclear criticality would be permitted as long as the dose thresholds of § 70.61(b) and § 70.61(c) are not exceeded.

Thus, 10 CFR Part 70 contains three separate and distinct requirements related to precluding nuclear criticality (10 CFR 70.64(a)(9), 10 CFR 70.61(b), and 10 CFR 70.61(d)), besides provisions in § 70.24 and § 70.52, which pertain to mitigating the consequences of a criticality accident and reporting its occurrence.

Section 70.61(d) of 10 CFR Part 70

Section 70.61(d) requires that under normal and credible abnormal conditions all nuclear processes are subcritical including use of an approved margin of subcriticality for safety. In addition, preventive controls and measures must be the primary means of protection against criticality. Meeting this performance requirement entails a number of factors. First, all normal and credible abnormal conditions must be identified. There are many different methods that may be employed to do this, but a systematic methodology should be used to provide reasonable assurance that the complete spectrum of credible conditions has been identified.

Normal conditions are those specifically allowed for as part of the normal modes of operation in the facility design (i.e., conditions that may occur without the failure of any items relied on for safety (IROFS)). Abnormal conditions are those events not planned for as a regular occurrence in the facility or operation design. They include those undesirable conditions that are the result of external events and process deviations, including those resulting from the failure of identified IROFS. Credible abnormal events include both credible single events (e.g., an external event or failure of a single IROFS) and credible sequences of events. Credible sequences of events include, but may not be limited to, chains of independent but not unlikely process deviations (i.e., not precluded by IROFS) and chains of related failures of IROFS (i.e., failures that are not independent). Some judgment must be employed in determining what constitutes a credible abnormal condition. It is not necessary to include multiple independent failures of IROFS within the spectrum of credible abnormal conditions. Additional guidance on what is considered not credible is contained in NUREG-1520, Section 3.4.3.2:

a.  "An external event for which the frequency of occurrence can conservatively be estimated as less than once in a million years."

b.  "A process deviation that consists of a sequence of many unlikely human actions or errors for which there is no reason or motive...."

c.  "Process deviations for which there is a convincing argument, given physical laws, that they are not possible, or are unquestionably extremely unlikely...."

The requirement that nuclear processes be subcritical is satisfied if the licensee or applicant demonstrates that the most reactive credible conditions are subcritical. To provide adequate assurance of subcriticality, this must include margin. There are several different ways to demonstrate sub-criticality, as discussed below:

- If subcriticality is demonstrated using an appropriately validated calculation method, then $k_{eff}$ (K effective) (including calculation's uncertainties) must be less than the approved upper subcritical limit (USL), as specified in the license. Meeting this requires that models bound actual anticipated conditions (e.g., tolerances and uncertainties appropriately taken into account, most reactive credible system parameters allowed are assumed), as specified in the license. Additional guidance is provided in the criticality chapter of NUREG-1520, "Standard Review Plan for the Review of a License Application for a Fuel Cycle Facility," (Sections 5.4.3.4.1, 5.4.3.4.2, and 5.4.3.4.4).

- Subcritical margin may also be expressed in terms of system parameters rather than system $k_{eff}$. An example would be where the licensee or applicant has committed to use mass or dimensional limits that are some specified fraction of the critical values of those parameters. In such cases, the approach used must be approved by the NRC.

- Subcriticality may be demonstrated on the basis of subcritical limits included in the license, U.S. Nuclear Regulatory Commission (NRC) endorsed American National Standards Institute (ANSI) standards, or other documents that have been approved or endorsed by NRC. Approval or endorsement by the NRC implies that the Agency has found these references to include an acceptable margin of subcriticality for safety.

- Industry handbooks of criticality data may also be used if widely accepted in the nuclear industry and if used in accordance with any limitations of that data. The NRC, however, reserves the right to evaluate the use of such handbooks on a case-by-case basis.

The requirement that preventive controls and measures be the primary means of protection against criticality is satisfied if engineered or administrative controls relied on to meet § 70.61(d) are designed to prevent occurrence of the critical excursion rather than mitigate its consequences. By stating that prevention should be the *primary* means of protection, it is recognized that there may be extraordinarily rare occasions when prevention alone is not sufficient to meet § 70.61(d). Such cases require convincing demonstration that there is no practicable way to meet § 70.61(d) with solely preventive measures.

Some examples where the § 70.61(d) requirement has not been met:

- A process in which the most reactive credible conditions have not been modeled and have not been shown to have $k_{eff}$ less than the approved USL.

- A process in which subcriticality is based on criticality calculations, but the model is outside the area of applicability of the calculation method.

- A process for which there is an unanalyzed or unanticipated credible abnormal condition (e.g., unanticipated failure of an IROFS or unanticipated external event).

- A process for which there is a credible common-mode event that can result in the failure of all criticality controls such that it can lead to a critical configuration.

- A process in which the designated IROFS are not sufficient to limit the system to a subcritical configuration.

<u>Relationship of 10 CFR 70.61(b) to 10 CFR 70.61(d)</u>

Section 70.61(b) states ". . . the risk of each credible high consequence event must be limited. . . . Controls . . . shall be applied to the extent needed to reduce the likelihood of occurrence of the event so that . . . the event is highly unlikely . . . ."

Section 70.61(d) states ". . . the risk of nuclear criticality accidents must be limited by assuring that under normal and credible abnormal conditions, all nuclear processes are subcritical, including an approved margin of subcriticality . . . ."

As written, the rule language requires <u>both</u> provisions (i.e., § 70.61(b) and § 70.61(d)) be met, since § 70.61(d) states "In addition to complying with paragraphs (b) and (c) of this section . . . ." However, during the 10 CFR Part 70 rulemaking, regulated industry representatives met with NRC and submitted letters in which they expressed their desire that NRC not consider criticality accidents high consequence events and not associate quantitative likelihoods with double contingency. As discussed in the release notes issued with the 10 CFR Part 70 rulemaking, in response to industry arguments accidental criticality was explicitly removed from the high consequence (§ 70.61(b)) category and a separate performance requirement for criticality (§ 70.61(d)) was created. The staff felt that in so doing, both the industry's desires as well as the staff's needs would be met. Further, the staff felt that the § 70.61(d) requirement required the same information as that required by § 70.61(b). Saying all nuclear processes must be subcritical in § 70.61(d) implies that criticality events must be prevented. Moreover, since likelihood is never zero, some non-zero likelihood must be assumed; the highly unlikely requirement in § 70.61(b) is appropriate for this. Therefore, the staff felt that by removing criticality explicitly from § 70.61(b) and creating § 70.61(d) during the rulemaking, the staff still retained its desired outcome–to prevent criticality accidents from occurring. The final rule Statement of Considerations (SOC) stated that "...the NRC believes that a separate performance requirement for nuclear criticality prevention is appropriate. The staff recognizes that many (but not all) nuclear criticality accidents would reasonably be expected to result in worker doses that exceed the high- and intermediate-consequence standards in 10 CFR 70.61(b) or (c). However, regardless of the dose directly resulting from the accident, an inadvertent nuclear criticality should be avoided. This is consistent with the Commission's goal to prevent inadvertent criticalities, as reflected in the NRC Strategic Plan (NUREG-1614) . . . ." However, there remained ambiguity regarding the relationship between § 70.61(b) and § 70.61(d). While the staff's intent was to have a single performance requirement for criticality accidents, this cannot be substantiated by a literal examination of the final rule.

Comparing the language in § 70.61(b) and (d), one concludes that § 70.61(d) is actually more restrictive than § 70.61(b). Section 70.61(d) essentially requires that there be no criticality accidents, with a high degree of assurance, whereas § 70.61(b) essentially requires that deaths and injuries (as implemented through a dose limit) be precluded (i.e., be made to be highly unlikely). If criticality accidents are prevented, then deaths and injuries are also prevented. However, the converse is not necessarily true; if deaths and injuries are prevented, criticality accidents are not necessarily prevented. Therefore, if one meets § 70.61(d), then one also automatically meets § 70.61(b); and if one meets § 70.61(b) through preventive means, and also meets the additional requirements specified in § 70.61(d), then one also meets § 70.61(d) in full. Thus, if a licensee chooses to address criticality event sequences under 10 CFR 70.61(b) with a preventive strategy and has an approved margin of subcriticality for safety, then the licensee will have also met the requirements under 10 CFR 70.61(d). However, if the licensee chooses to address criticality event sequences under 10 CFR 70.61(b) with a mitigative strategy, then the licensee will not have met the requirements under 10 CFR 70.61(d) and additional controls will have to be identified to ensure subcriticality.

Another consideration is that both § 70.61(b) and § 70.61(d) set the standard that must be met (i.e., the performance requirements), but not the methodology. Methodology requirements are contained in § 70.62. One cannot look at § 70.61 in a vacuum. All other 10 CFR Part 70 provisions must also be met, including the § 70.62(c) provision that requires the integrated safety analysis (ISA) to include radiological hazards, facility hazards, potential accident sequences, and identification of IROFS as well as the assumptions and conditions under which the IROFS are relied upon to support compliance with § 70.61 performance requirements. It also requires that the ISA team include a person with experience in criticality safety. These requirements must be met regardless of whether the licensee attempts to meet the performance requirements starting from § 70.61(b) or § 70.61(d). The three options below can be seen to be equivalent when one considers that § 70.62 must also be met for all cases.

To meet the regulations and prevent criticalities, an applicant/licensee may use one of the three approaches below (in conjunction with other 10 CFR Part 70 requirements, including those in § 70.62):

1.    Demonstrate compliance with § 70.61(d); or
2.    Demonstrate compliance with § 70.61(b), considering only preventive controls and including an approved margin of subcriticality; or
3.    Separately demonstrate compliance with both § 70.61(d) and § 70.61(b).

Use of any of the above three approaches will satisfy the regulations.

That both § 70.61(b) and § 70.61(d) apply to criticality is supported by this SRP. In addition, there are several references to the requirement to make criticality highly unlikely.

Double Contingency Principle § 70.64(a)(9)

In addition to complying with the performance requirement in § 70.61, new facilities and processes are required to comply with the baseline design criteria in § 70.64. Section 70.64(a)(9) requires that the design provide for criticality control, including adherence to the double contingency principle (DCP). In addition to this requirement for new facilities and processes, many existing facilities and processes have license commitments to meet the DCP for licensed activities. Although Subpart H of 10 CFR Part 70 is relatively new, this conceptual framework is not new. Licensees have historically committed to ANSI/American Nuclear Society -8.1 (ANSI/ANS-8.1). This standard also requires that nuclear processes be ensured to be subcritical under normal and credible abnormal conditions. By contrast, the DCP is stated as a recommendation of ANSI/ANS-8.1. Therefore, the standard recognizes that adherence to the DCP can be one means, but is not necessarily the only means of meeting the underlying subcriticality requirement. The conditions under which compliance with the DCP ensures that § 70.61(d) is met are discussed below.

The double contingency principle is a design principle intended to be used in designing a facility that meets the performance requirements of § 70.61. The definition in § 70.4 ("...process designs *should* incorporate sufficient factors of safety...") implicitly recognizes that there may be some cases in which a strict adherence to the double contingency principle is not practicable. This should be an exceedingly rare situation and should be accompanied by a convincing demonstration that a strict adherence to the double contingency principle is not practicable. Section 70.64(a) allows for this in stating that licensees must maintain the application of this criterion unless the integrated safety analysis (ISA) demonstrates that it is not relied on for safety or otherwise does not require adherence.

The presence of two controls may not be necessary, or may not be sufficient, to meet the DCP. The DCP does not necessarily require two controls; it requires "at least two...changes in process conditions" be needed before criticality is possible. Meeting this may necessitate one, two, or more than two controls depending on the possible conditions that can lead to criticality. In general, there will be many pathways to criticality and, therefore, more than two controls required to meet the DCP for an entire process.

In addition, § 70.64(b)(1) requires that the design must incorporate, whenever practicable, preference for the selection of engineered over administrative controls to increase overall system reliability. Passive engineered controls are generally preferable to active engineered controls, and engineered to administrative controls. In addition, process design should rely on geometry control as opposed to control of other parameters whenever practicable, and on diverse means of control (e.g., reliance on two different criticality parameters or different means of controlling one parameter) whenever practicable, to minimize the potential for common-mode failure. Cases in which these preferences cannot be complied with will generally require more justification to show adherence with the DCP. For example, one cannot claim that the double contingency principle is met with only two controls (regardless of type) if the resulting configuration fails to protect against all credible pathways to criticality or limit the risk of inadvertent criticality as required in 10 CFR 70.61(d).


Relationship between § 70.61 and § 70.64(a)(9)

As stated above, adherence to the DCP can be one means of meeting the performance requirements of § 70.61(d) (and, therefore, also § 70.61(b)). Historically, a number of different approaches to double contingency have been used. Some cases that have been used in the past may not be sufficiently robust to satisfy the performance requirements of § 70.61. Typically, this has been due to a reliance on controls that were not sufficiently robust (e.g., weak administrative controls). The purpose of this guidance is not to promote a new standard for all applications but rather to clarify when adherence to the DCP will establish a sufficient basis for meeting the performance requirements. To facilitate this, the following guidance is provided on the various terms in the definition of the DCP:

Unlikely changes in process conditions should be expected to occur rarely, or not at all, during the lifetime of the facility. Operational events that occur regularly should not be credited as a contingency relied on to meet the DCP (although they may constitute part of a contingency if a combination of events may be considered unlikely). Therefore, the occurrence of any such event generally reveals a deficiency in the design that should result in corrective action.

Determination that a contingency is unlikely should be based on objective attributes of the criticality controls, rather than on subjective judgment alone. Examples of such attributes are environmental factors that can degrade the reliability and availability of controls, margin, and redundancy and diversity of controls. (Guidance on some of the availability and reliability qualities that should be considered is provided in Section 3.4.3.2(9) of this SRP and NUREG-1718, "Standard Review Plan for the Review of a License Application for a Mixed Oxide (MOX) Fuel Fabrication Facility," Section 5.4.3.2(B)(vii).) Management measures should be provided, as needed, to ensure that the failure of the criticality controls is an unlikely contingency. (NOTE: Usage of the term "unlikely" in the DCP is not equivalent to the term as used in § 70.61(c) for intermediate consequence events.)

Independent changes in process conditions are such that one contingency neither causes another contingency nor increases its likelihood of occurrence. The existence of any credible common-mode failure of both contingencies means that it is not valid to consider them independent. For example, related actions performed by the same individual or using the same equipment will not generally be sufficiently independent to meet the DCP.

Concurrent does not mean that the two changes in process conditions must occur simultaneously, but that the effect of the first contingency persists until the second contingency occurs. Prompt detection and correction of abnormal conditions should thus be provided to restore double contingency protection. The time required to detect and correct failures should be significantly shorter than the anticipated time between failures in order for there to be significant risk reduction provided from failure detection.

Changes in process conditions does not imply that reliance on two different parameters is mandatory to meet the DCP. Reliance on two different parameters is preferable to reliance on two controls on a single parameter, however, because of the difficulty in achieving complete independence when controlling one parameter. In those cases in which single parameter control is unavoidable, great care should be taken to ensure that no common-mode failures exist.

In addition to meeting the above, the following guidance is provided to illustrate the conditions under which adherence to the double contingency principle (in terms of the guidance above) is sufficient to meet the performance requirement of 10 CFR 70.61:

- Controls are established on system parameters to preclude changes in process conditions, and these controls are designated as IROFS in accordance with § 70.61(e). (Reliance should be based on items that are designated as IROFS in the ISA Summary and not on random factors that may or may not be maintained.)

- The condition resulting from the failure of a leg of double contingency has been shown to be subcritical with an acceptable margin (e.g., $k_{eff}$ is less than USL, parameters are within subcritical limits specified in the license or endorsed standards).

- Controls are sufficiently reliable to ensure that each change in process conditions necessary for criticality is "unlikely." Management measures are established to ensure that they are available and reliable to perform their safety function.

Because the DCP is only one means of meeting the performance requirements, it is possible to meet the DCP without meeting the conditions above (including designating criticality controls as IROFS in the ISA Summary). In this case, however, another method must be relied on to meet the § 70.61 performance requirements. However, in order to use compliance with the DCP as part of the demonstration of meeting the § 70.61 performance requirements, these conditions should be met.

Some specific examples of control systems that meet § 70.61(d) through use of the DCP follow:

*A passive geometry control in which no credible failure mode (e.g., bulging, corrosion, or leakage) exists and which has been placed under configuration management:*

- A favorable geometry vessel in a benign environment in which corrosion or other material degradation is not credible. In addition, the vessel is of such robust construction (e.g., thick stainless steel, steel surrounded by concrete) that it is unquestionably not going to leak, and there is no credible mechanism for the material to accumulate in an unfavorable configuration.

- A tank that is not authorized to contain fissile material is located far outside the fissile material handling areas and is physically isolated from fissile liquid processes by a blank flange or siphon break, such that backflow is not credible.

*Two passive controls in which there is a credible failure mode, and there are sufficient management measures to ensure the controls continue to perform their safety functions (e.g., periodic surveillance to detect corrosion/bulging):*

- A favorable geometry solution column, in which leakage of the tank is a credible upset. In addition, the column is in an area in which the solution would leak into a favorable geometry dike, and the leakage would be self-revealing (i.e., column is in a continually manned area) or the column and dike would be subject to periodic surveillance.

- A double-sleeved solution line in which leakage of the inner pipe would be quickly detected (e.g., by conductivity probe between the pipes or by transparent baffling).

- A storage array in which fissile material is stored in fixed geometry containers, and the spacing between containers is fixed by birdcages or other fixed devices, and geometry and spacing controls are ensured by configuration management and periodic walkthroughs.

*One passive control under configuration management and one active engineered control whose reliability is ensured by periodic functional testing, maintenance, and an alarm to automatically indicate its failure:*

- A calciner relying on geometry and moderation control in which geometry control is provided by limiting the calciner interior to the height of a single layer of pellet boats, and moderation control is provided by monitoring of the calciner temperature. Temperature control is ensured by thermocouples that alarm if the temperature drops below a minimum set-point.

A down-blending tank that is subcritical for uranium solutions with less than a limiting enrichment in which volume control is provided by the design of the tank and enrichment control is provided using mass flow totalizers and a mechanical stirrer. The failure of these active devices automatically stops the transfer of solution and actuates an alarm.

A large geometry tank relying on raschig rings for criticality control in which the raschig rings are only approved up to a limiting concentration, and the concentration is controlled by an in-line sodium iodide detector that closes an isolation valve when actuated.

*One engineered and one enhanced administrative control in which the instrumentation and devices included in the administrative control are subject to periodic functional testing and maintenance, and the operator action is performed routinely or reinforced by periodic drills and training:*

A powder handling glovebox relying on moderator and mass control in which moderator control is provided by the glovebox design (e.g., airtight, dry nitrogen atmosphere, sloped ventilation ductwork) and mass is procedurally controlled by limiting batch size. In addition, mass transfers must be logged into a computer tracking system that alarms if mass limits are exceeded.

A vessel in which the volume of fissile solution is controlled by the diameter of the tank and by procedurally limiting the solution height. In addition, operator actions are backed up with a high-level switch equipped with an alarm.

*One engineered control and one simple administrative control in which the reliability of the administrative control is subject to a high degree of redundancy:*

Solution transfer from favorable to unfavorable geometry relying on two controls on concentration. Two different operators are required to draw separate samples which are then analyzed in the laboratory by two different methods and shown to be within concentration limits before transfer is authorized. In addition, the area supervisor maintains control of a key to the transfer pump so that the procedure may not be inadvertently bypassed. This is backed up with an in-line sodium iodide detector that automatically closes an isolation valve if concentration limits are exceeded.

(NOTE: Use of two independent samples is generally not considered adequate for both legs of double contingency because of the difficulty in ensuring complete independence between the samples.)

*Two administrative controls that are independent (e.g., performed by different individuals or verified by a supervisor), for which human factors have been considered in the design of the process such that the operation is not prone to error, and there is sufficient margin to require multiple failures before the criticality control limit can be exceeded:*

A glovebox relying on dual mass control in which two operators or an operator and a supervisor must confirm that placing material into the glovebox will not result in the mass limit being exceeded. In addition, criticality would require the mass limit to be exceeded multiple times, which would be difficult to achieve and would be readily apparent.

- A drum storage array limited to a vertical stack of four drums in which there are no forklifts in the area capable of raising a drum above this height. In addition, the drums are very heavy and violating the stack height limit would require an immense physical effort.

- A planar storage array in which mass-controlled containers are procedurally limited to not less than 24 inches center-to-center, and in which criticality would require assembling a very large number of containers into a spherical heap and reflecting them intimately with water.

*Other considerations ensuring that there is no credible event leading to criticality:*

- A facility handling uranium enriched to no more than 1 weight percent (wt%) uranium-235 ($^{235}$U).

- A facility in which the site-wide limit is less than a minimum critical mass.

- A facility storing contaminated soil or equipment with a very low uranium concentration in which there is no known concentration mechanism that can lead to a critical configuration.

Some examples of control systems that would not meet § 70.61(d) through use of the DCP:

*Double contingency consisting of two single operator actions without any supervisor verification or redundancy:*

- Solution transfers that rely only on two operators drawing separate samples or in which a single procedural deviation could cause an unauthorized transfer.

- A mass controlled system in which triple batching (i.e., two successive batching errors) could result in criticality when the mass transfers are done by a single operator.

- A storage array in which two violations on administrative spacing requirements could credibly lead to criticality.

*A leg of double contingency consisting of an administrative control for which correct performance of the action cannot be readily confirmed or is subjective:*

- A solution vessel in which the operator is required to confirm concentration or chemical form by visually observing a color change in the solution.

- A tank in which the operator is required to verify prior to operation that the tank is "essentially empty."

*A leg of double contingency consisting of complex administrative tasks composed of multiple steps that are susceptible to error:*

- A glovebox in which the operator is required to calculate the mass of plastic, paper, and other miscellaneous materials in order to comply with moderator control.

- A solution transfer operation in which one leg consists solely on a single sample being correctly drawn, labeled, analyzed, recorded, and read.

- Maintenance on a dissolution process in which criticality safety relies on the correct performance of a procedure to replace an in-line filter. The procedure requires that the filter be removed, flushed, and re-installed in a multi-step process that has several opportunities for failure.

*A leg of double contingency consisting of an administrative control with insufficient margin to ensure that the safety limit will not be exceeded:*

- A glovebox in which mass is controlled administratively, and in which the normal mass limit is almost equal to the minimum critical mass.

- A planar storage array in which spacing between containers is administratively limited to be less than 24 inches center-to-center, and in which criticality will result if a few containers are placed 23 inches apart.

*A leg of double contingency consisting of an engineered control in which there is no reasonable means to detect and correct the failure within a given time.*

- A solution process in which it is plausible for concentrated solution to be allowed to accumulate undetected over a long period of time in an unfavorable geometry.

- A vessel in which geometry control is provided by a double wall, but there is no means of detecting leakage between the walls. In addition, the vessel is of a type known to have a history of leakage (e.g., heat exchanger).

*A leg of double contingency consisting of a control in an environment where its safety function is degraded.*

- A solution vessel relied on for geometry control, but which is subject to pressure fluctuations that can cause the vessel to bulge beyond a favorable diameter.

- Instrumentation whose performance is degraded under conditions that can be reasonably expected during normal operations (e.g., temperature, pressure, presence of corrosive gases, or loss of essential utilities such as electricity, plant air, or water).

*A leg of double contingency consisting of a control where its behavior under adverse conditions is uncertain.*

- An unfavorable geometry pump in which mass control relies on the presumption that the pump will malfunction before an unsafe volume of uranium accumulates in the pump oil, and for which no failures of this type have been observed.

*A leg of double contingency consisting of undeclared design features or process conditions that are not precluded by being explicitly controlled.*

A powder blending operation in which uranium oxide density that is less than the theoretical density is assumed, but the process variables affecting density (e.g., calcinations temperature, mechanical pressure of the pellet press) are not specifically controlled and there is no confirmatory sampling.

A solvent extraction process in which nominal concentration of uranyl nitrate is assumed, but there is no in-line monitoring or confirmatory sampling.

A vault in which the mass limit is not controlled by procedure or license limit, but is merely based on current inventory.

A process relying on the favorable geometry of passive equipment, but for which the dimensions and/or material composition are not specifically identified as criticality controls.

This list is merely illustrative and not meant to be exhaustive. However, these examples demonstrate that double contingency that satisfies the performance requirements can be based on one, two, or more than two passive engineered, active engineered, or administrative controls, and that reliability and availability of those controls depends on management measures, safety margins, environmental conditions, human factors, and other process and control characteristics. Not every application similar to these examples will be found acceptable—determination must be made on the totality of the information available, and an analyst should consider all factors that may degrade the robustness of the controls.


**Technical Review Guidance**

Relationship of 10 CFR 70.61(b) and 10 CFR 70.61(d)

The reviewer needs to assure that all applicable 10 CFR Part 70 criticality provisions (including § 70.62(c)) are met. To meet the regulations and prevent criticalities an applicant/licensee may use one of the three approaches below (in conjunction with other 10 CFR Part 70 requirements, including those in § 70.62):

1.    Demonstrate compliance with § 70.61(d); or
1.    Demonstrate compliance with § 70.61(b), considering only preventive controls and including an approved margin of subcriticality; or
2.    Separately demonstrate compliance with both § 70.61(d) and § 70.61(b).

Use of any of the above three approaches will satisfy the regulations.

Staff should not dictate which of the above three options must be met; rather, staff should assure that the applicant/licensee has met one of these options.

Double Contingency Principle
One way, but not the only way, of meeting 10 CFR 70.61 is by applying the double contingency principle (defined in 10 CFR 70.4) to accident sequences leading to criticality that are required to be developed per § 70.62. Adherence to the DCP will satisfy the performance requirement of § 70.61(d) (and therefore also § 70.61(b)) provided the following conditions are met:

- Controls are established on system parameters to preclude changes in process conditions, and these controls are designated as IROFS in accordance with § 70.61(e). (Reliance should be based on items that are designated as IROFS in the ISA Summary and not on random factors that may or may not be maintained.)

- The condition resulting from the failure of a leg of double contingency has been shown to be subcritical with an acceptable margin (e.g., $k_{eff}$ is less than USL, parameters are within subcritical limits specified in the license or endorsed standards).

- Controls are sufficiently reliable to ensure that each change in process conditions necessary for criticality is "unlikely." Management measures are established to ensure that they are available and reliable to perform their safety function.

In the absence of meeting these conditions, an alternate demonstration of compliance with the performance requirements should be provided.

# 6. CHEMICAL PROCESS SAFETY

## 6.1 Purpose of Review

The primary purpose of the review is to determine, with reasonable assurance, that the applicant has designed a facility that will provide adequate protection against chemical hazards related to the storage, handling, and processing of licensed materials, as required by Title 10 of the *Code of Federal Regulations* (10 CFR) Part 70, "Domestic Licensing of Special Nuclear Material." The facility design must adequately protect the health and safety of workers and the public from the chemical risks in the facility during normal operations and credible accident conditions. It must also protect against facility conditions that could affect the safety of licensed materials and thus present an increased radiation or chemical risk (e.g., a chemical that incapacitates operators and prevents their entry into an area of the facility where licensed materials are handled). As defined in 10 CFR 70.4, "Definitions," a hazardous chemical is one produced from licensed material (e.g., hydrogen fluoride, which is produced by the reaction of uranium hexafluoride and water), or one that is in contact with licensed material, and that is toxic, explosive, flammable, corrosive, or reactive to the extent that it can endanger life or health if not adequately controlled. The U.S. Nuclear Regulatory Commission (NRC) does not regulate substances before process addition to licensed material or after process separation from licensed material, unless the chemical could potentially contact licensed materials (e.g., by leaks or human error) or cause an increased radiological risk.

Chemical safety issues are initially evaluated as part of the applicant's integrated safety analysis (ISA) summary. As required in 10 CFR 70.65, "Additional Contents of Applications", the ISA Summary must evaluate credible accident sequences at the facility, identify items relied on for safety (IROFS) to prevent the occurrence or to mitigate the consequences of accidents, and include the management measures that provide reasonable assurance of the availability and reliability of IROFS when needed. To begin the chemical safety review, the reviewer should examine the license application, the facility and process description in Chapter 1 of this Standard Review Plan (SRP), and the ISA Summary (SRP Chapter 3) to gain familiarity with the following:

- process information and accident sequences leading to conditions that could pose chemical hazards

- IROFS and sole IROFS[1] used to prevent or mitigate such chemical hazards

- proposed procedures to protect public health and safety and the environment (e.g., a high-level programmatic description of how the licensee or applicant proposes to operate, maintain, or manage the facility)

- the definition of "unlikely," "highly unlikely," and "credible," as used in the ISA evaluations

- quantitative standards used to assess the consequences to an individual from acute chemical exposure to licensed material or a chemical produced from licensed materials that are on site or expected to be on site

---

- [1] Sole IROFS are those that are the sole item preventing or mitigating an accident for which the consequences could exceed the performance requirements of 10 CFR 70.61.

- recommended management measures for ensuring that the IROFS will be available and reliable when required

## 6.2 Responsibility for Review

Primary:     Chemical Process Safety Reviewer (all sections of this chapter)

Supporting:  Licensing Project Manager
             Fuel Cycle Facility Inspection Staff (as needed)
             Health Physicist (for uranium and transuranic toxicity issues)
             Primary Reviewers of Chapters 1, 3, 8, 9, and 11 of this SRP

## 6.3 Areas of Review

Regulations in 10 CFR 70.62(a) require an applicant to establish and maintain a safety program that will adequately protect worker and public health and safety and the environment from the chemical hazards from licensed material. Although not necessarily required to establish a separate chemical process safety program, the applicant must demonstrate that it has considered chemical hazards and accident sequences that affect licensed material and has adequately prevented or mitigated them in accordance with 10 CFR 70.61, "Performance Requirements." Applicants must conduct an ISA and provide an ISA Summary that meets the requirements of 10 CFR 70.65, "Additional Content of Applications."

The staff's chemical process safety review should focus on the chemical safety-related accident sequences described in the ISA Summary (SRP Chapter 3) and the corresponding management measures (SRP Chapter 11) to confirm that the applicant's equipment, facilities, and management measures are adequate to protect against releases and chemical exposures of licensed material, hazardous chemicals produced from licensed material, and chemical risks of plant conditions that affect the safety of licensed material. The review must verify that any grading of IROFS or management measures proposed by the applicant in accordance with 10 CFR 70.62(a) is commensurate with the accident risk that the IROFS are designed to reduce.

The 1988 memorandum of understanding between the NRC and the Occupational Safety and Health Administration (OSHA) directs the NRC to oversee chemical safety issues related to (1) radiation risks of licensed materials, (2) chemical risks of licensed materials, and (3) plant conditions that affect or may affect the safety of licensed materials or the availability or reliability of IROFS and thus increase radiation risk to workers, the public, and the environment. The NRC does not oversee plant conditions that do not affect or involve the safety of licensed materials.

The staff's review should cover the following specifications:

- chemical process description, providing a narrative description of the site, facility, and processes with respect to chemical safety for normal operations, which can include process chemistry, flow diagrams, major process steps, and major pieces of equipment; the ISA summary must include a reasonably simple description of each process (unit operations)

- chemical accident sequences, including unmitigated accident sequences involving hazardous chemicals and licensed materials and interpretation of the quantitative chemical risk levels, as described in the ISA Summary

- chemical accident consequences, identified in the ISA Summary, including the applicant's interpretation of the qualitative chemical risk levels and the assumptions, bases, and methods the applicant used to forecast the consequences to workers and the public of accidents that involve hazardous chemicals and licensed materials

- chemical process IROFS and sole IROFS, including a list of items relied on for chemical safety and a description of their safety function, as described in the ISA Summary

- chemical process management measures, including management measures to ensure the reliability and availability of IROFS (chemical process safety), as described in the ISA Summary

Review Interfaces

In addition to Chapter 6 of the application, the chemical reviewer should examine information in the following other areas to ensure that it is consistent with Chapter 6:

- facility and process description applied to chemical safety, as described in Chapter 1 of this SRP

- safety program, ISA commitments, and ISA documentation applied to chemical safety under SRP Chapter 3

- emergency plan applied to chemical safety under SRP Chapter 8

- dispersion models used for consequence modeling under SRP Chapter 9

- configuration management, maintenance, training and qualifications, procedures, audits and assessments, incident investigations, record management, and other quality assurance elements, as described in SRP Chapter 11

## 6.4 Acceptance Criteria

### 6.4.1 Regulatory Requirements

Acceptance criteria are based on meeting the relevant requirements of the following regulations:

- The general and additional contents of an application for chemical process safety are in 10 CFR 70.22, "Contents of Applications," and 10 CFR 70.65, respectively. General information that must be included in the license application appears in 10 CFR 70.22. Information that must be included in the ISA Summary appears in 10 CFR 70.65.

- The requirements for the approval of the application are in 10 CFR 70.23, "Requirements for the Approval of Applications."

- The chemical process safety review should provide reasonable assurance of compliance with the performance requirements in 10 CFR 70.61.

- Requirements to maintain and establish a safety program appear in 10 CFR 70.62, "Safety Program and Integrated Safety Analysis"

- Requirements for new facilities or new processes at existing facilities that require a license amendment under 10 CFR 70.72, "Facility Changes and Change Process," appear in 10 CFR 70.64, "Requirements for New Facilities or New Processes at Existing Facilities"

## 6.4.2 Regulatory Guidance

The following regulatory guidance is relevant to chemical process safety:

- NUREG-1513, "Integrated Safety Analysis Guidance Document," May 2001

- NUREG-1391, "Chemical Toxicity of Uranium Hexafluoride Compared to Acute Effects of Radiation," February 1991

- NUREG-1601, "Chemical Process Safety at Fuel Cycle Facilities," August 1997

- NUREG/CR-6410, "Nuclear Fuel Cycle Facility Accident Analysis Handbook," March 1998

- NUREG/CR-6481, "Review of Models Used for Determining Consequences of $UF_6$ Release," November 1997

## 6.4.3 Regulatory Acceptance Criteria

The reviewer should find the applicant's chemical process safety information acceptable if there is reasonable assurance that it adequately addresses and satisfies the acceptance criteria presented below. The applicant may elect to incorporate some or all of the requested chemical process information in the facility and process description (SRP Section 1.1) or the ISA Summary, rather than in this section. Either approach is acceptable, as long as the information is adequately cross-referenced.

### 6.4.3.1 Chemical Process Description

The regulation in 10 CFR 70.65(b)(3) requires the ISA Summary to include a description of each process in the facility. The applicant's descriptions of the chemical processes are acceptable if they meet the following conditions:

- Process descriptions are sufficiently detailed to allow an understanding of the chemical process hazards (including radiological hazards caused by or involving chemical accidents) and to allow the development of potential accident sequences.

- Process descriptions are sufficiently detailed to allow an understanding of the theory of operation.

### 6.4.3.2 Chemical Accident Sequences

The chemical accident sequences are acceptable in the following circumstances:

- The applicant provides a general description of the accident sequences identified in the ISA Summary as involving hazardous chemicals produced from licensed material or chemical risks of plant conditions that affect the safety of licensed materials.

- The ISA Summary describes the chemical hazards identified in the ISA that could result in a high- or intermediate-consequence event. Each accident sequence identified by the applicant in the ISA should include a chemical hazard evaluation of potential interactions of process chemicals with confinement vessels, process equipment, and facility personnel. The hazard evaluation should use appropriate accepted methods.

- The applicant provides reasonable assurance that measures to mitigate the consequences of accident sequences identified in the ISA Summary are consistent with actions described in SRP Chapter 8. (Note that some facilities are not required to have an emergency plan.)

### 6.4.3.3 Chemical Accident Consequences

The chemical accident consequences are acceptable if the following apply:

- The applicant identifies and uses appropriate techniques and valid assumptions in estimating the concentrations for releases of hazardous chemicals produced from licensed material or by abnormal plant conditions that could affect the safety of licensed materials.

- The applicant provides evidence that the dispersion models used to determine whether a release of chemicals might affect worker or public health and safety are appropriate to the process. The applicant should demonstrate that the models used lead to a conservative estimate of potential consequences.[2]

- Consequence analyses conform to the guidance on atmospheric and consequence modeling in NUREG/CR-6410.

- If the applicant does not use the methods in NUREG/CR-6410, it may propose an alternative method accompanied by supporting documentation to justify the selection of such an alternative.

- The application describes the quantitative standards (chemical concentration limits) used to assess the unmitigated and mitigated consequences to an individual outside the control area (public) or to a worker from acute chemical exposure to licensed material, chemicals produced from licensed materials, or chemicals in contact with licensed materials that are on site or expected to be on site.

---

[2]    Source term and vapor dispersion models used to calculate the concentration of uranium hexafluoride ($UF_6$) and its reaction products conform to the guidance on the applicability of models in NUREG/CR-6481.

- Acceptable exposure standards include, but are not limited to, the Emergency Response Planning Guidelines established by the American Industrial Hygiene Association, the Acute Exposure Guideline Levels established by the National Advisory Committee for Acute Guideline Levels for Hazardous Substances, and the exposure limits established by OSHA. The applicant needs to verify that the selected standard applies to the worker or the individual outside the control area. Note that all the standards mentioned above apply to airborne exposure to gases, vapors, and particulates. Those limits are not intended to evaluate consequences for chemical exposures through other exposure paths.

- If the applicant does not use a published exposure standard or knows of no such standard for a chemical, it may propose an alternative, accompanied by supporting documentation to justify it.[3]

To ensure that the performance requirements in 10 CFR 70.61 are not exceeded, the applicant's ISA should evaluate the degree of hazard and routes of entry of the hazardous chemicals. Material safety data sheets contain useful information about toxicity, health effects, first aid, reactivity, protective equipment, and spill or leak procedures. Therefore, it is highly recommended that the licensee review the material safety data sheets of the hazardous chemicals at the facility when accessing chemical consequences. The reviewer should confirm the following:

- Consequence categorization is in accordance with the performance requirements in 10 CFR 70.61(b) and 10 CFR 70.61(c).

- The application includes definitions of "unlikely," "highly unlikely," and "credible," as used in the evaluations in the ISA.

### 6.4.3.4 Chemical Process IROFS and Sole IROFS

The license application should identify the design basis for chemical process safety for normal operation and demonstrate that the proposed equipment and facilities adequately protect public health and safety and the environment. Based on a comparison of the unmitigated chemical consequences determined in Section 6.4.3.3 with the performance criteria of 10 CFR 70.61, the applicant must provide (in the ISA Summary) a list of chemical process safety controls (i.e., IROFS) suitable to prevent or mitigate potential accidents. This list should also briefly describe the IROFS, in sufficient detail to permit an understanding of their safety functions. The application should identify IROFS for those accident sequences containing a chemical system or process failure that may ultimately lead to radiological consequences that exceed the performance requirements. In order to comply with the requirements in 10 CFR 70.61, the applicant must demonstrate that the likelihood of each credible high-consequence event will be reduced after the implementation of IROFS, so that the event will be highly unlikely or the consequence of the event will be low. For each credible intermediate-consequence event, after the implementation of controls, the event should be unlikely or its consequences should be low.

---

[3] Note that 10 CFR 70.61 requirements are for "acute chemical exposures," and OSHA permissible exposure limits are typically time-weighted average values. Consequently, for ISA purposes, acute chemical release limits may not be adjusted by the time-weighted average calculation (which involves concentration and duration of exposure) unless the ISA Summary provides a rational basis.

If the applicant takes a graded approach to safety, in accordance with 10 CFR 70.62(a), the reviewer should establish that the grading of IROFS is appropriate and sufficient to protect against chemical process risks. For example, the applicant should consider reliance on passive controls of active systems and defense in depth, in accordance with 10 CFR 70.64(b). To reduce common-mode failures, the applicant should favor design features that use independent sources of motive force for items such as control actuators, jet pumps, eductors, and ejectors. Fail-safe controls are preferred unless safety concerns prevent this approach.

### 6.4.3.5 Chemical Process Management Measures

The applicant should review management measures to ensure the availability and reliability of IROFS and sole IROFS when they are required to perform their safety functions. Management measures may be graded, commensurate with risk.

The application should meet the following criteria:

- The application should describe the engineering approach, basis, or schemes employed to maintain safety during normal operations.

- The ISA Summary should identify the administrative and engineered controls to prevent or mitigate a chemical process risk, the hazard being mitigated, and the risk category. The applicant should also explain how it graded the safety of IROFS and management measures and how such grading is commensurate with the reduction in risk that the IROFS are designed to achieve.

- The application should demonstrate that the management measures ensure that IROFS are available and reliable by briefly describing the following:

  - procedures to ensure the reliable operation of engineered controls (e.g., inspection and testing procedures and frequencies, calibration programs, functional tests, corrective and preventive maintenance programs, criteria for acceptable test results)

  - procedures to ensure that administrative controls will be correctly implemented, when required (e.g., employee training and qualification in operating procedures, refresher training, safe work practices, development of standard operating procedures, training program evaluation)

### 6.4.3.6 Requirements for New Facilities or New Processes at Existing Facilities

The application should address the baseline design criteria (BDC) for new facilities or new processes at existing facilities that require a license amendment under 10 CFR 70.72. The applicant must apply the BDC to the design of new processes (§ 70.64 (a)(5)) but is not required to retrofit existing facilities or existing processes; however, all facilities and processes must comply with the performance requirements in 10 CFR 70.61. NUREG-1601, Section 2.4, contains a list of items that an adequate facility design should consider. For new facilities and processes in existing facilities, the design must provide for adequate protection against chemical risk from licensed material, facility conditions that affect the safety of licensed material, and hazardous chemicals produced from licensed material. With respect to chemical process safety, the application should be considered acceptable if it includes the following information (or references other sections of the application that include this information):

- The applicant briefly describes how it performed the ISA for the new process and how the ISA satisfies the principles of the BDC and the performance requirements in 10 CFR 70.61. The applicant also explains how it applies defense in depth to higher risk accident sequences. Acceptable defense-in-depth principles for the chemical process safety design are those that support a hierarchy of controls: prevention, mitigation, and operator intervention, in order of preference.

- The applicant describes proposed facility-specific or process-specific relaxations or additions to BDC, along with justifications for relaxations.

- The ISA Summary describes how the applicant applied the chemical safety BDC in establishing the design principles, features, and control systems of the new process.

## 6.5  Review Procedures

### 6.5.1  Acceptance Review

During the acceptance review of a license application, the reviewer should scan the submittals to identify major deficiencies in the information provided for each area of review specified in SRP Section 6.3. Reviewers must decide whether they have enough information to proceed with a detailed review. Obvious problems that can be addressed in a single request for additional information should be accepted. However, before the NRC performs a detailed review, the applicant should correct major deficiencies that would require several requests for additional information to resolve.

Reviewers should record whether each area of review is adequately addressed in the application, is adequately addressed in a referenced document, is not applicable to the application, or has a major deficiency.

### 6.5.2  Safety Evaluation

During the safety evaluation, the reviewer determines whether the application comprehensively describes the chemical safety of the licensed activity, as identified in SRP Section 6.3. For deviations from the specific acceptance criteria, the staff should review the applicant's explanation of how the proposed alternatives to the SRP criteria provide an acceptable method of complying with the relevant NRC requirements identified in Section 6.4.

During the license application and ISA Summary review, the reviewer should identify and note any items or issues that should be inspected during an operational readiness review, if such a review will be performed. These items could include confirming that the applicant implemented engineered controls through procedures and operator training.

For an existing facility, the reviewer may consult NRC inspectors to identify and resolve any issues related to the licensing review. For a planned facility, the reviewers may wish to consult with the facility design team to gain a better understanding of the process, its potential hazards, and its safety approaches. The reviewers should coordinate these interactions through the licensing project manager.

The primary reviewer will prepare input to the safety evaluation report (SER) for the licensing project manager in support of the licensing action.

### 6.5.2.1 Chemical Process Description

The results of the ISA are the basis for the chemical process safety evaluation. The reviewer should establish that the applicant's facility design, operations, and IROFS for chemical safety provide reasonable assurance that they will function as intended and ensure the safe handling of licensed material at the facility. The reviewer must verify that the applicant's proposed equipment and facilities are adequate to protect public health and safety and the environment. The reviewer should examine the mechanisms that will allow the applicant to identify and correct potential problems.

### 6.5.2.2 Chemical Accident Sequences

The ISA shall contain the potential accident sequences caused by process deviations or other events internal to the facility and credible external events, including natural phenomena. The reviewer should assess the chemical risks identified in the ISA Summary and ensure that the design and the operational plans for the facility reflect the level of safety. In return, to validate the criteria used by the applicant in reporting sequences in the ISA Summary, the reviewer will make an independent judgment of the comparative risks assigned by the applicant to accident sequences identified in the ISA Summary. This judgment is based on risk relative to other sequences (competing risks), the complexity of the sequence, facility operating history, and general industry performance. Whenever possible, a licensee should use its own experience to supplement the identification of potential chemical hazards. The review may cover a selected number of lower risk, chemical safety-related accident sequences not identified in the ISA Summary.

### 6.5.2.3 Chemical Accident Consequences

The reviewer must verify that the proposed quantitative standards used to assess the consequences to an individual from acute chemical exposure are appropriate. Events with high and intermediate consequences should be identified, and IROFS should be used to reduce the likelihood or the consequences of the event. The reviewer needs to ensure that the select standards are correctly applied to the worker or the member of the public.

### 6.5.2.4 Chemical Process IROFS and Sole IROFS

The staff reviews the chemical process safety IROFS to ensure their adequacy in protecting against all unmitigated sequences identified in the ISA Summary. The reviewer should establish that the applicant's controls (IROFS) for chemical safety provide reasonable assurance that they will function as intended and ensure the safe handling of licensed material at the facility.

### 6.5.2.5 Chemical Process Management Measures

The application should include a chemical safety element describing the methods, activities, and implementation of the overall safety program. The technical reviewer should verify the applicant's commitment to retaining records for chemical process safety compliance and reporting commitments for chemical releases. In addition, the reviewer should verify the applicant's commitment to refer any unacceptable performance deficiency to those responsible for the facility's corrective action function, in accordance with Chapter 11 of this SRP.

If the applicant has applied a graded approach to safety, the reviewer should establish that the grading of IROFS or management measures is appropriate and sufficient to protect against chemical process risks (see Chapter 11 of this SRP).

### 6.5.2.6 *Requirements for New Facilities or New Processes at Existing Facilities*

The staff reviews the applicant's commitments to adhere to the BDC, according to 10 CFR 70.64(a), for the design of new facilities or new processes at an existing facility that require a license amendment under 10 CFR 70.72.

## 6.6 Evaluation Findings

The reviewer writes an SER input addressing each topic reviewed and explains why the NRC staff has reasonable assurance that the chemical safety portion of the application is acceptable. The reviewer may propose license conditions to impose requirements where the application is deficient. If unable to make a finding of reasonable assurance, the reviewer will prepare SER input explaining the deficiencies and the reasons for denying the proposed application. In cases where the SER is drafted in advance of resolving all outstanding chemical process safety issues, the reviewer documents the review as described below and includes a list of open issues that require resolution before the staff can make a finding of reasonable assurance. For partial reviews, revisions, and process changes, the reviewer uses applicable sections of the acceptance criteria and the SER, and the reviewer notes areas that were not reviewed and their chemical process safety significance, if any. On completion of the review, the NRC staff may impose temporary license conditions to authorize short-duration activities. For certain functions and requirements that concern safety or regulatory issues, the NRC staff may impose a license condition that remains in effect until removed by an amendment or license renewal.

The SER should include a summary statement of what the NRC staff evaluated and the basis for the reviewer's conclusions. The SER should include statements like the following:

> The staff has evaluated the application using the criteria listed previously. Based on the review of the license application, the NRC staff has concluded that the applicant has adequately described and assessed accident consequences that could result from the handling, storage, or processing of licensed materials and that could have potentially significant chemical consequences and effects. The applicant has constructed a hazard analysis that identified and evaluated those chemical process hazards and potential accidents and established safety controls to provide reasonable assurance of safe facility operation. To ensure that the performance requirements in 10 CFR Part 70 are met, the applicant has provided reasonable assurance that controls are available and reliable when required to perform their safety functions. The staff has reviewed these safety controls and the applicant's plan for managing chemical process safety and finds them acceptable.

> The staff concludes that both the applicant's plan for managing chemical process safety and the chemical process safety controls meet the requirements of 10 CFR Part 70 and provide reasonable assurance that the health and safety of the public will be protected.

## 6.7 References

Center for Chemical Process Safety, "Guidelines for the Technical Management of Chemical Process Safety," American Institute of Chemical Engineers, New York, 1989, Chapter 11, as revised.

Chemical Manufacturers Association, "Responsible Care®, Process Safety Code of Management Practices," Washington, DC, 1990.

*U.S. Code of Federal Regulations*, Chapter I, Title 10, "Energy," Part 70, "Domestic Licensing of Special Nuclear Material."

*U.S. Code of Federal Regulations*, Title 29, "Labor," Section 1910.100, Chapter XVII, "Occupational Safety and Health Administration: Subpart Z—Toxic and Hazardous Substances, Tables Z-1 and Z-2."

U.S. Nuclear Regulatory Commission, "Inspection of the Nuclear Chemical Process Safety Program at Fuel Cycle Facilities," Inspection Manual Chapter 2603, as revised.

U.S. Nuclear Regulatory Commission/Occupational Safety and Health Administration, "Memorandum of Understanding between the Nuclear Regulatory Commission and the Occupational Safety and Health Administration, 'Worker Protection at NRC-Licensed Facilities,'" *Federal Register*, Vol. 53, October 31, 1988.

U.S. Nuclear Regulatory Commission, "Chemical Process Safety at Fuel Cycle Facilities," NUREG-1601, August 1997.

U.S. Nuclear Regulatory Commission, "Integrated Safety Analysis Guidance Document," NUREG-1513, May 2001.

U.S. Nuclear Regulatory Commission, "Chemical Toxicity of Uranium Hexafluoride Compared to Acute Effects of Radiation," NUREG-1391, February 1991.

U.S. Nuclear Regulatory Commission, "Nuclear Fuel Cycle Facility Accident Analysis Handbook," NUREG/CR-6410, March 1998.

U.S. Nuclear Regulatory Commission, "Review of Models Used for Determining Consequences of $UF_6$ Release," NUREG/CR-6481, November 1997.

American Industrial Hygiene Association, "Emergency Response Planning Guidelines," as revised. The latest available list of approved guidelines, dated January 1, 2008, is available on the home page of the Subcommittee on Consequence Assessment and Protective Actions (http://orise.orau.gov/emi/scapa/default.htm).

National Research Council, "Acute Exposure Guidelines Levels for Selected Airborne Chemicals," National Academy of Sciences (ISBN: 978-0-309-12755-4), Volume 7, Washington DC, 2008.

# 7. FIRE SAFETY

## 7.1 Purpose of Review

The purpose of this review is to determine with reasonable assurance that the applicant has designed a facility that provides adequate protection against fires and explosions that could affect the safety of licensed materials and thus present an increased radiological or chemical risk. The review should also establish that the applicant has considered the radiological and chemical consequences of the fires and will institute suitable safety controls to protect workers, the public, and the environment.

Fire safety issues are initially evaluated as part of the applicant's integrated safety analysis (ISA) summary. The ISA Summary must evaluate credible accident sequences at the facility; identify items relied on for safety (IROFS) to prevent the occurrence or to mitigate the consequences of accidents; and include the management measures that provide reasonable assurance of the availability and reliability of IROFS, when needed. Reviewers assess the applicant's approach to protecting against fire and explosion hazards by examining the license application and the ISA Summary to gain familiarity with the following:

- process information and accident sequences leading to conditions that could pose fire hazards

- IROFS and sole IROFS used to prevent or mitigate such fire hazards

- management measures applied to ensure that IROFS will be available and reliable when required

## 7.2 Responsibility for Review

Primary:    Fire Safety Specialist

Secondary:  Criticality Safety Specialist
            Environmental Specialist
            Chemical Safety Specialist
            Physical Security Specialist

Supporting: Regional, Resident, and Fuel Cycle Inspection Staff

## 7.3 Areas of Review

Title 10 of the *Code of Federal Regulations* (10 CFR) 70.62(a) requires an applicant to develop, implement, and maintain a safety program that will reasonably protect public health and safety and the environment from the fire and explosive hazards associated with processing, handling, and storing licensed materials during normal operations, anticipated operational occurrences, and credible accidents. The fire protection program must address these process-specific risks and general fire prevention, protection, and management issues. Although 10 CFR Part 70, "Domestic Licensing of Special Nuclear Material," does not require a separate fire safety program, an applicant should provide commitments pertaining to fire safety in the following areas:                \

- Fire safety management includes safety organization, engineering review, and fire prevention; inspection, testing, and maintenance; prefire plans; and personnel qualifications, drills, and training.

- Fire risk identification includes the fire hazards analysis (FHA) and the ISA Summary.

- Facility design includes information on building construction, fire areas, life safety, ventilation, and electrical system design. The facility design should also consider competing requirements among fire safety and security, criticality, and environmental concerns.

- Process fire safety involves design considerations to prevent an accident or to mitigate the consequences of an accident resulting from the use of process chemicals, combustible metals, flammable and combustible liquids and gases, high-temperature equipment, hot cells and glove boxes, and laboratories.

- Fire protection systems include fire detection, alarm, and suppression systems; portable extinguishers; water supplies; and emergency response organizations.

Review Interfaces

In addition to Chapter 7 of the application, the reviewer should examine information in the following other areas to ensure that it is consistent with the information in Chapter 7:

- Review information about the facility and process descriptions related to fire safety as required under Chapter 1 of this Standard Review Plan (SRP).

- Review information on the safety program, ISA commitments, and ISA documentation applied to fire safety as required under SRP Chapter 3.

- Review information on controls applied to chemical processes for fire safety as required under SRP Chapter 6.

- Review information on configuration management, maintenance, training and qualifications, procedures, audits and assessments, incident investigations, record management, and other quality assurance (QA) elements as required under SRP Chapter 11 as related to fire safety.

## 7.4 Acceptance Criteria

An applicant that meets the acceptance criteria defined in this section or that has provided an acceptable alternative should be considered as having provided reasonable assurance of an acceptable fire protection program.

### 7.4.1 Regulatory Requirements

The regulatory basis for the fire safety review should be the requirements of 10 CFR 70.22, "Contents of Applications," and 10 CFR 70.65, "Additional Content of Applications." In addition, the fire safety review should focus on providing reasonable assurance of compliance with 10 CFR 70.61, "Performance Requirements"; 10 CFR 70.62, "Safety Program and Integrated

Safety Analysis"; and 10 CFR 70.64, "Requirements for New Facilities or New Processes at Existing Facilities."

## 7.4.2 Regulatory Guidance

The relevant regulatory guidance for fire safety includes the following U.S. Nuclear Regulatory Commission (NRC) and industrial standards:

- National Fire Protection Association (NFPA) 801, "Standard for Fire Protection for Facilities Handling Radioactive Materials," latest edition

- NUREG-1513, "Integrated Safety Analysis Guidance Document," May 2001

- NUREG/CR-6410, "Nuclear Fuel Cycle Facility Accident Analysis Handbook," March 1998

## 7.4.3 Regulatory Acceptance Criteria

Partial acceptability of the application and the ISA Summary will be contingent on the NRC staff's review of the applicant's commitments to control and mitigate fire hazards. The staff will focus on whether the application is risk informed, addresses the applicant's procedures for maintaining an acceptable level of fire safety, and demonstrates that the applicant is prepared to react quickly and safely to extinguish fires. An applicant may use a graded approach to define fire safety, but it must provide sufficient documentation and commitments to ensure that it will adequately protect workers, the public, and the environment from fire events.

The applicant may incorporate these acceptance criteria in the information supplied to satisfy SRP Chapter 3 (regarding ISA) or other SRP chapters as long as it provides clear cross-references (information need not be repeated). The staff's fire safety specialist will review the application, ISA Summary, and other documentation, as needed, regarding these acceptance criteria.

The reviewer(s) will use nationally recognized codes and standards, as appropriate, in evaluating a reasonable assurance of fire safety. These codes and standards include, but are not limited to, the NFPA "National Fire Codes"; Factory Mutual Research Corporation data sheets and approval guide; Underwriters Laboratories, Inc., standards and building material directory; American National Standards Institute standards; and American Society for Testing and Materials standards. Commitments to specified standards will normally be considered an acceptable means of meeting the acceptance criteria.

The NRC staff will review the application to ensure that it meets the acceptance criteria discussed below.

### 7.4.3.1 Fire Safety Management Measures

An adequate application documents how the applicant will administer and ensure fire safety at the licensed facility. The application should reflect a commitment to ensure that the IROFS, as identified in the ISA Summary, are available and reliable and that the facility maintains fire safety awareness among employees, controls transient ignition sources and combustibles, and maintains a readiness to extinguish the fire or limit its consequences. These measures are unique to fire safety and, therefore, are not included in the acceptance criteria for SRP Chapter 11.

An adequate application identifies a senior-level manager who has the authority and staff to ensure that fire safety receives appropriate priority. A facility safety committee or fire safety review committee staffed by managers of different disciplines should integrate facility modifications. (The facility safety committee can do the work of a fire safety review committee.) As described in the application, an individual with sufficient practical fire safety experience in nuclear facilities should supervise day-to-day fire safety.

NFPA 801 specifies the following fire safety management measures:

- fire prevention
- inspection, testing, and maintenance of fire protection systems
- emergency response organization qualifications, drills, and training
- prefire plans

An adequate application documents the fire safety management measures in sufficient detail to identify their relationship to, and functions in, normal operations, anticipated (off-normal) events, and accident safety (i.e., IROFS). The staff recognizes NFPA 801 as one acceptable standard for fire safety management measures; however, the applicant may use other nationally recognized codes and standards if appropriate.

### 7.4.3.2 Fire Hazards Analysis

7.4.3.2.1 Development of a Fire Hazard Analysis as a Tool for Evaluating Fire Hazards

Knowing the fire risk allows an applicant to apply the appropriate level of fire protection to ensure the safety of workers, the public, and the environment from fire-induced radiological or chemical hazards. To be risk informed, a licensee should conduct an FHA for each facility, or part thereof, that, if totally consumed by fire, could release special nuclear material (SNM) in quantity and form that could cause at least an intermediate consequence, as defined in 10 CFR 70.61. The FHA should develop bounding credible fire scenarios for each fire area containing significant fire loading and then assess the consequences of an unmitigated fire. The staff recognizes NFPA 801 as one standard that provides guidance for conducting FHAs; however, the applicant may use other nationally recognized codes and standards if appropriate. The FHA should include a description, by fire area, of the fuel loading, fire scenarios, methods of consequence analysis, the potential consequences, and a description of the mitigative or preventive controls or both.

The FHA is used to identify possible fire initiators and accident sequences leading to radiological consequences or toxic chemical consequences resulting from interaction with SNM. In developing accident sequences that will be reported in the ISA Summary, the ISA team will consider the FHA results and assign likelihoods to the various events in the accident sequences. With respect to fire safety, the ISA Summary is acceptable if it identifies the credible fire hazards (e.g., from the FHA) for each process fire and if it provides details as to how the applicant considered and addressed (i.e., the management measures and IROFS) each fire hazard for each process accident sequence whose consequence could exceed the performance requirements in 10 CFR 70.61. Thus, the FHA is a fundamental tool for evaluating fire hazards as input to the ISA evaluation.

### 7.4.3.2.2 Deviations from National Fire Protection Association Codes and Standards

When the applicant or licensee states that its design "meets the NFPA code(s)" or "meets the intent of the NFPA codes" and does not identify any deviations from such codes, the NRC expects that the design conforms to the codes and is subject to inspection against the NFPA code of record. A design that "meets the intent of the code" should specify those sections of the code with which it does not conform. A licensee may apply the equivalency concept in meeting the provisions of the NFPA codes or standards. Nothing in the NFPA codes or standards is intended to prevent the use of methods, systems, or devices of equivalent or superior quality, strength, fire resistance, durability, and safety as alternatives to those prescribed by the codes or standards, provided that technical documentation demonstrates equivalency and that the method, system, or device is listed or approved for the intended purpose. Recent editions of the NFPA codes require submittal of technical documentation to the "authority having jurisdiction" to demonstrate equivalency of an alternative system, method, or device. The NRC does not require review and approval of equivalency evaluations before their implementation during construction. However, the licensee should document these evaluations and make them available for NRC inspection. The NRC recognizes that fire protection systems and controls may be required to meet State or local codes and may need to be approved by code enforcement officials. Where such systems and controls are not required to meet the performance requirements of 10 CFR 70.61 (i.e., not designated as IROFS), a State or local code enforcement official may be designated as the authority having jurisdiction (as described in NFPA documents). However, the NRC must review and inspect IROFS and any code deviations relative to their effect on nuclear safety. The authority having jurisdiction refers to the NRC Director of the Office of Nuclear Material Safety and Safeguards (or designee).

### 7.4.3.3 Facility Design

Building construction, fire area determination, electrical installation, life safety, ventilation, drainage, and lightning protection are all facility design features that affect fire safety. The staff recognizes NFPA 801 as one standard that specifies acceptable facility fire safety design criteria; however, the applicant may use other nationally recognized codes and standards, if appropriate. An adequate application documents the fire safety considerations used in the general design of fuel cycle facilities. The NRC normally reviews the following design information related to fire safety in a license application:

- the type of construction (as required under NFPA 220, "Standard on Types of Building Construction," for a new building) and applicable building codes (for an existing building) with a comparison to NFPA 220 building types

- the identification of building material, the fire duration rating (if known), and a description of exterior openings

- the overall description of the fire detection system, including the degree of compliance with NFPA 72, "National Fire Alarm and Signaling Code," for design, installation, surveillance, testing, and maintenance

- the overall description of the automatic fire suppression system, applicable design standards, the system design basis, identification of standards for surveillance, testing, and maintenance procedures

- the description of the water distribution system, including descriptions of fire pumps, fire mains, the location of sectionalizing valves, maximum fire demand, and compliance with applicable NFPA standards

In addition to standard industrial fire safety concerns, the application should also address the following nuclear safety, environmental protection, and physical security issues:

- Criticality concerns may exclude water extinguishing systems from process areas. However, during major fire events, the fire may easily overcome the extinguishing capability of portable extinguishers, and hose lines may be needed to extinguish the fire. Consequently, applicants should consider using total flooding gaseous systems in water-exclusion areas with significant fire risks. An adequate application addresses the methodology for extinguishing fires in water-exclusion areas.

- Environmental concerns include the potential for thousands of gallons of fire water to be contaminated with nuclear material during a fire event. Consequently, diked areas and drainage of process facilities may be needed. NFPA 801 provides guidance on how to calculate the potential amount of runoff to properly size drainage and containment systems. An adequate application documents any measures used to control fire water runoff.

- Physical security concerns include the need to design buildings and facilities to provide safe egress in case of fire. Physical security requirements for SNM may inadvertently delay worker egress and firefighter access. Physical security procedures should allow offsite fire departments quick and efficient access to fire emergencies. An adequate application documents the design criteria used for worker egress and procedures for firefighter access. The staff recognizes NFPA 801 as one standard that specifies acceptable worker egress design criteria; however, the applicant may use other nationally recognized codes and standards, if appropriate.

Design and construction of new facilities must comply with the baseline design criteria (BDC) specified in 10 CFR 70.64(a) and comply with the defense-in-depth requirements of 10 CFR 70.64(b). The design and construction should be consistent with the guidance provided in NFPA 801 or other appropriate nationally recognized fire protection codes and standards.

### 7.4.3.4 Process Fire Safety

Many hazardous chemicals and processes used by fuel cycle facilities contribute to the fire hazards. In areas that have fire hazards that may threaten licensed material, the application should identify the hazardous chemicals, processes, and design standards used to ensure fire safety. The staff recognizes NFPA 801 as one standard that provides acceptable design criteria for radiological process areas that may contain hazardous material, laboratories, high-temperature equipment, hot cells, and glove boxes. However, the applicant may use other nationally recognized codes and standards, if appropriate.

The following are a few of the more common hazardous materials used at fuel cycle facilities:

- Anhydrous ammonia is an explosive, flammable, and toxic gas used to make hydrogen.

- Fluorine reacts violently with organic material or metal powders and water vapor.

- Hydrogen is an explosive and flammable gas used in reduction processes.

- Hydrogen peroxide off-gases hydrogen and oxygen and is incompatible with some extinguishers.

- Nitric acid nitrates organic material, which lowers the ignition temperature of combustibles.

- Sulfuric acid absorbs water from organic material in an exothermic reaction, thereby causing ignition.

- Zirconium is a combustible metal that burns at elevated temperatures.

- Calciners and incinerators are sources of heat that have initiated fires at fuel cycle facilities.

The applicant should identify fire and explosion hazards, fire and explosion parameters of hazardous materials, and the degree of compliance with applicable codes (e.g., NFPA 30, "Flammable and Combustible Liquids Code"; NFPA 69, "Standard on Explosion Prevention Systems"; and NFPA 86, "Standard for Ovens and Furnaces").

In addition to participating in the integrated review of the ISA Summary performed in accordance with Chapter 3 of this SRP, the reviewer should also examine in detail the fire-initiated release scenarios provided in the ISA Summary to demonstrate compliance with 10 CFR 70.61. This review should follow the guidance provided in applicable subsections of SRP Chapter 3 to include a detailed evaluation of these scenarios, including a review of fire initiators, fire-induced consequences, the likelihoods of such consequences, and IROFS chosen to prevent or mitigate those consequences.

7.4.3.4.1 Fire-Initiated Accident Sequences

The review should consider the following factors in determining the acceptability of the applicant's descriptions of fire-initiated accident sequences:

- The applicant provided enough detail in its fire hazard descriptions to permit an understanding of the fire hazards sufficient to allow an evaluation of potential accident sequences.

- The applicant adequately described the consequences and likelihoods of accident sequences identified in the ISA Summary involving fire, including risks from hazardous chemicals produced from licensed material and risks from radioactive materials.

- The applicant provided enough detail in its justification of the initiation probability for the reviewer to make an independent verification for those scenarios in which the initiation probability appears to be unconservative. If a facility relies on controls to achieve this initiation probability, the applicant should identify these controls as IROFS, as appropriate.

- Controls that are used to mitigate or prevent the scenario are identified as IROFS or as defense-in-depth measures. For those controls that are IROFS, reliability and associated management measures must be indicated.

- Analyses that the applicant has performed as part of the evaluation are part of the ISA and referenced.

7.4.3.4.2 Items Relied on for Safety and the Associated Management Measures

Based on a comparison of the unmitigated fire protection accident sequence consequences with the performance criteria of 10 CFR 70.61, the applicant should identify (in the ISA Summary) fire protection safety controls suitable to prevent or mitigate potential accidents. If the applicant takes a graded approach to safety in accordance with 10 CFR 70.62(a), the reviewer should establish that the grading of IROFS and associated management measures are appropriate and sufficient to protect against fire-related risks. A minimum acceptable level of requirements for management measures (e.g., QA for fire protection systems) is the level that NFPA codes require, such as the use of equipment listed by an acceptable organization (e.g., Underwriters Laboratories, Inc., or Factory Mutual Global). Installation and initial testing should also be that specified by the appropriate code.

The NRC staff should also review those management measures that ensure the availability and reliability of such IROFS when they are required to perform safety functions. The ISA Summary should demonstrate that the management measures proposed to ensure that IROFS are available and reliable when required by briefly describing the following:

- measures to ensure the reliable operation of engineered controls (e.g., inspection and testing procedures and frequencies, calibration programs, functional tests, corrective and preventive maintenance programs, and criteria for acceptable test results)

- measures to ensure that administrative controls will be correctly implemented when required (e.g., employee training and qualification in operating procedures, refresher training, safe work practices, development of standard operating procedures, and training program evaluations)

- the compliance of IROFS with all applicable NFPA or industry consensus fire codes and standards

At minimum, IROFS should comply with those sections of the codes or standards affecting the reliability and effectiveness of the IROFS. For example, fire protection systems do not need to be seismically designed beyond what is required by the applicable NFPA code if they are not intended to function during a seismic event.

7.4.3.5 Fire Protection and Emergency Response

The application should document the fire protection systems and fire emergency response organizations provided for licensed facilities. The ISA Summary should identify the fire protection IROFS. An adequate application describes the fire protection for areas in which licensed material is present. The application should describe which standards the fire protection systems and equipment meet. The staff recognizes NFPA's national fire codes as acceptable standards for the design, installation, testing, and maintenance of the fire protection systems and equipment. However, the applicant may use other nationally recognized codes and standards, if appropriate.

Facilities with the potential for rapidly developing fires that do not have an adequate nearby emergency responder may need an onsite fire emergency response team. One acceptable standard is NFPA 600, "Standard on Industrial Fire Brigades." However, the applicant may use other nationally recognized codes and standards, if appropriate. If offsite fire departments are needed for facility fire safety, periodic training with the fire departments is necessary so that offsite departments will become familiar with facility access procedures, facility layout, and prefire plans. A memorandum of understanding between the applicant and the fire departments is recommended to define the required protection. The staff's fire safety specialist will review the adequacy of the applicant's fire protection and emergency response commitments.

### 7.4.3.6 Requirements for New Facilities or New Processes at Existing Facilities

The application or ISA Summary or both should address the BDC as required under 10 CFR 70.64 for new facilities or new processes at existing facilities that require a license amendment under 10 CFR 70.72, "Facility Changes and Change Process." With respect to fire safety, the application should be considered acceptable if it includes the information listed below (or references other sections of the application that include this information):

- The application should briefly describe how the ISA was performed for the new process, including its use and relationship to the performance requirements in 10 CFR 70.61, the BDC, and a defense-in-depth strategy for higher risk accident sequences. Acceptable principles for defense in depth of the fire safety design would be those that support the hierarchy of controls with preference for prevention of releases (over mitigation of consequences) and engineered controls over administrative controls.

- The ISA Summary should describe how the applicant applied 10 CFR 70.64(a)(3) in establishing the design principles, features, and control systems of the new process. This will normally involve a commitment to follow appropriate codes and standards for design, testing, surveillance, and maintenance of fire protection systems, including those that are not IROFS but may involve nuclear processes or buildings housing nuclear material.

## 7.5 Review Procedures

### 7.5.1 Acceptance Review

During the acceptance review, the primary reviewer evaluates the application for completeness as required by 10 CFR Part 70 and determines whether the application addresses the criteria discussed in SRP Section 7.3. If significant deficiencies in the application are identified, the primary reviewer should recommend the return of the application or request additional information before the start of the safety evaluation.

### 7.5.2 Safety Evaluation

During the safety evaluation, the primary and secondary reviewers determine whether the application comprehensively describes the fire safety of the licensed activity as identified in SRP Section 7.3 and assess the commitments made in response to the criteria specified in SRP Section 7.4. The staff may ask the applicant or licensee to provide additional information or modify the submittal to meet the acceptance criteria.

Reviewers should note that NFPA 801 uses "administrative control" in a different sense than how the term is used in 10 CFR Part 70 and elsewhere in this SRP. In 10 CFR Part 70, an administrative control is an IROFS if it is the human action necessary to meet safety performance requirements and if it is supported by management measures (e.g., training, QA, and procedures) that ensure that the action will be taken if needed. In NFPA 801, "administrative controls" refer to the training, qualifications, and procedures behind the human action; however, these elements are referred to as "management measures" in 10 CFR Part 70 and in this SRP.

For an existing facility, the reviewer may consult cognizant NRC inspectors to identify and resolve any issues related to the licensing review. For a planned facility, the reviewers may wish to consult with the facility design team to gain a better understanding of the process, its potential hazards, and safety approaches.

The primary reviewer will prepare a safety evaluation report (SER) for the licensing project manager in support of the licensing action.

#### 7.5.2.1 Fire-Related Risks and Accident Sequences

The results of the ISA are the basis for the fire safety evaluation. The reviewer should assess the fire risks identified in the ISA Summary and ensure that the level of safety is reflected in the design and the operational plans for the facility. The reviewer should establish that the applicant's facility design, operations, and IROFS for fire and explosion safety provide reasonable assurance that they will function as intended and will provide for the safe handling of licensed material at the facility.

#### 7.5.2.2 Items Relied on for Safety and Management Measures

The staff reviews the fire and explosion IROFS to ensure their adequacy in protecting against all unmitigated sequences identified in the ISA Summary.

If the applicant has applied a graded approach to safety, the reviewer should establish that the grading of IROFS or management measures is appropriate and sufficient to protect against fire and explosion risks.

#### 7.5.2.3 Requirements for New Facilities or New Processes at Existing Facilities

The staff reviews the applicant's commitments as required to satisfy the BDC in 10 CFR 70.64(a) for the design of new facilities or new processes at an existing facility that require a license amendment under 10 CFR 70.72.

## 7.6 Evaluation Findings

The staff's review should verify that the applicant (1) provides sufficient information to satisfy the intent of 10 CFR Part 70 requirements related to the overall safety program and (2) is consistent with the fire safety criteria in this SRP. On the basis of this information, the staff should be able to evaluate whether the application meets the appropriate criteria. As an example, the staff might document the fire safety review in an SER in the following manner:

- The applicant has established a fire protection function meeting the acceptance criteria in SRP Chapter 7. The function includes a facility safety review committee responsible for integrating modifications to the facility and a fire safety manager responsible for day-to-day program implementation. Fire prevention, inspection, testing, and maintenance of fire protection systems and the qualification, drills, and training of facility personnel are in accordance with applicable NFPA codes and standards. (Note that SER Section 11.3 describes fire protection training requirements.)

- The applicant has conducted risk analyses in accordance with NFPA 801. The FHAs identified credible fire scenarios that bound the fire risk. The ISA used these scenarios and identified fire protection IROFS (in particular, wet pipe sprinkling in the process areas, isolation of the high-temperature equipment within fire barriers, and a fire brigade meeting NFPA 600). A memorandum of understanding with the fire department documents the required assistance and the annual exercises. Procedures are in place to allow the fire department efficient access to process areas during fire emergencies. Worker egress is designed and maintained in accordance with NFPA 101, "Life Safety Code."

- The applicant has demonstrated that it incorporated appropriate fire safety considerations in the design of its facilities. The applicant has also demonstrated that the facility has appropriate active fire protection systems.

- The staff concludes that the applicant's submittals provide sufficient information in accordance with requirements of 10 CFR 30.33 and 10 CFR 40.32, both entitled "General Requirements for Issuance of Specific Licenses," and those of 10 CFR 70.22 and 10 CFR 70.65 on potential fire hazards, consequences, and required controls for the proposed processes. The NRC staff determined that the applicant demonstrated compliance with the performance requirements of 10 CFR 70.61 for fire protection related to postulated accident scenarios. The design that the applicant proposes also satisfies the requirements of 10 CFR 70.64(a)(3) and the defense-in-depth requirements of 10 CFR 70.64(b) (as required).

## 7.7 References

*U.S. Code of Federal Regulations*, Chapter I, Title 10, "Energy," Part 70, "Domestic Licensing of Special Nuclear Material."

National Fire Protection Association (NFPA), "National Fire Codes," Standard 30, "Flamable and Combustible Liquids Code," 2008.

National Fire Protection Association (NFPA), "National Fire Codes," Standard 69, "Standards of Explosion Prevention Systems," 2008.

National Fire Protection Association (NFPA), "National Fire Codes," Standard 72, "National Fire Alarm Code," 2010.

National Fire Protection Association (NFPA), "National Fire Codes," Standard 86, "Standards for Ovens and Furnaces." 1999.

National Fire Protection Association (NFPA), "National Fire Codes," Standard 101, "Life Safety Code," 2007.

National Fire Protection Association (NFPA), "National Fire Codes," Standard 220, "Standards of Types of Building Construction," 2009.

National Fire Protection Association (NFPA), "National Fire Codes," Standard 600, "Standards of Industrial Fire Brigades," 2010.

U.S. Nuclear Regulatory Commission, "Uranium Oxide Fires at Fuel Cycle Facilities," Information Notice 92-14, February 21, 1992.

U.S. Nuclear Regulatory Commission, "Integrated Safety Analysis Guidance Document," NUREG-1513, May 2001.

U.S. Nuclear Regulatory Commission, "Nuclear Fuel Cycle Facility Accident Analysis Handbook," NUREG/CR-6410, March 1998.

# 8. EMERGENCY MANAGEMENT

## 8.1 Purpose of Review

The purpose of reviewing the applicant's emergency management plan is to determine if the applicant has established, before the start of operations, adequate emergency management facilities and procedures to protect workers, the public, and the environment. In preparing its emergency plan, the applicant may use either this Standard Review Plan (SRP) or Regulatory Guide 3.67, "Standard Format and Content for Emergency Plans for Fuel Cycle and Materials Facilities," issued January 1992. The applicant may provide the information requested for the emergency plan once and then cross-reference it in other sections.

Regulations in 10 CFR 70.22, "Contents of Application" require an emergency management plan or an emergency evaluation if the licensee is authorized to possess (1) enriched uranium or plutonium for which a criticality accident alarm system is required, (2) uranium hexafluoride in excess of 50 kilograms (kg) (110 pounds (lb)) in a single container or a total of 1,000 kg (2,200 lb), or (3) plutonium in excess of 2 curies in unsealed form or on foils or plated sources. A licensed facility that meets the above criteria is required to possess an emergency management plan when an evaluation (or the integrated safety analysis (ISA) Summary referenced in lieu of the evaluation) shows that the maximum dose to a member of the public off site from a release of radioactive materials would exceed 0.01-sievert (Sv) (1-rem) effective dose equivalent or an intake of 2 milligrams (mg) of soluble uranium.

The baseline design criteria (BDC) of Title 10 of the *Code of Federal Regulations* (10 CFR) Part 70, "Domestic Licensing of Special Nuclear Material," incorporate emergency capability. The criteria are intended to ensure the control of licensed material, the evacuation of personnel, and the availability of emergency facilities.

## 8.2 Responsibility for Review

Primary:      Assigned Licensing Staff

Secondary:    Licensing Project Manager

Supporting:   Regional Emergency Preparedness Inspector
ISA Reviewer
Fuel Facility Inspection Staff

## 8.3 Areas of Review

The U.S. Nuclear Regulatory Commission (NRC) staff should review the applicant's submittal for an acceptable level of evidence of planning for emergency preparedness directed at situations involving real or potential radiological hazards. In particular, the review should address those design features, facilities, functions, and equipment that may affect some aspect of emergency planning or the capability of an applicant to cope with facility emergencies. In addition, the review should address coordination with offsite emergency response organizations. The staff should either review the emergency plan in accordance with 10 CFR 70.22(i)(1)(ii) and the guidance contained in the acceptance criteria below or review the applicant's evaluation (or the ISA Summary referenced in lieu of the evaluation) that an emergency plan is not needed in accordance with 10 CFR 70.22(i)(1)(i).

The NRC staff reviewer should address the areas of review, as described in Sections 8.3.1 and 8.3.2 below.

### 8.3.1 Emergency Plan

If the applicant submits an emergency plan, the staff should evaluate the emergency plan against 10 CFR 70.22(i)(1)(ii) and Regulatory Guide 3.67, which provides a standard format and content for an emergency plan. Elements in the emergency plan to be reviewed include the following:

- a facility description (including both onsite and offsite emergency facilities)

- the types of accidents

- the classification of accidents

- the detection of accidents

- the mitigation of consequences (and safe shutdown)

- the assessment of releases

- the responsibilities of the licensee

- notification and coordination

- information to be communicated and parties to be contacted

- training

- safe shutdown (recovery and facility restoration)

- exercises and drills

- hazardous chemical inventories and locations

- the responsibilities for developing and maintaining the emergency program and its procedures

### 8.3.2 Evaluation That No Emergency Plan Is Required

If the applicant submits an evaluation or references the ISA Summary to demonstrate that an emergency plan is not required, the staff should review the information against 10 CFR 70.22(i)(1)(i) and NUREG-1140, "A Regulatory Analysis on Emergency Preparedness for Fuel Cycle and Other Radioactive Material Licensees" dated January 1, 1988. NUREG/CR-6410, "Nuclear Fuel Cycle Facility Accident Analysis Handbook," issued March 1998, also contains useful information. Areas evaluated should include the following:

- a description of the facility

- the types of materials used, including both radioactive material and hazardous chemicals
- the types of accidents
- the detection of accidents
- site-specific information used to support the evaluation
- an evaluation of the consequences

Review Interfaces

In addition to Chapter 8 of the application, the reviewer should examine information in the following other areas to ensure that it is consistent with the information in Chapter 8:

- Review information about the facility, process description, geography, and demographics as applied to emergency planning under SRP Chapter 1.

- Review information on the safety program, ISA commitments, and ISA documentation applied to emergency planning under SRP Chapter 3.

- Review information about radiological releases under SRP Chapter 4.

- Review information about chemical releases under SRP Chapter 6.

- Review information on configuration management, maintenance, training and qualifications, procedures, audits and assessment, incident investigations, record management, and other quality assurance elements under SRP Chapter 11.

## 8.4 Acceptance Criteria

### 8.4.1 Regulatory Requirements

The regulation at 10 CFR 70.22(i)(1)(i) specifies when an applicant is not required to submit an emergency plan to the NRC; if an applicant is required to submit an emergency plan, 10 CFR 70.22(i)(3) describes the information that the emergency plan must include. In addition, 10 CFR 70.64(a)(6) requires applicants to address the control of licensed material, the evacuation of personnel, and the availability of emergency facilities for the design of new facilities.

### 8.4.2 Regulatory Guidance

Regulatory guidance for preparing an emergency plan includes the following sources:

- Regulatory Guide 3.67
- NUREG-1140
- NUREG/CR-6410

### 8.4.3 Regulatory Acceptance Criteria

*8.4.3.1 Emergency Plan*

The reviewer should evaluate the adequacy of the proposed emergency plan against the requirements in 10 CFR 70.22(i)(3) and the specific acceptance criteria provided in

Sections 8.4.3.1.1 through 8.4.3.1.14 of this SRP. The reviewer should find the applicant's emergency plan acceptable if it meets the regulatory requirements and the acceptance criteria described below.

8.4.3.1.1 Facility Description

The emergency plan should describe the facility and site, the area near the site, and the licensed activities. These descriptions should include the following:

- a detailed drawing of the site showing the following features:

    – onsite and near offsite (within 1.61 kilometers (km) (1 mile (mi))) structures with building numbers and labels

    – roads and parking lots on site and main roads near the site

    – site boundaries showing fences and gates

    – major site features

    – water bodies within approximately 1.61 km (1 mi)

- a general area map covering a radius of approximately 16.1 km (10 mi), a U.S. Geological Survey topographical quadrangle (7½-minute series, including the adjacent quadrangle(s) if the site is located less than 1.61 km (1 mi) from the edge of the quadrangle), and a map or aerial photograph indicating onsite and near-site structures within a radius of approximately 1.61 km (1 mi)[1]

- stack heights, typical stack flow rates, and efficiencies of any emission control devices

- a general description of licensed and other major activities conducted at the facility and the type, form, and quantities of radioactive and other hazardous materials that are normally on site, by location (use and storage) and building, and hazardous characteristics (exposure rates, pH, temperature, and other characteristics) that are important to emergency management

- certification by the plant manager (or the individual authorized by the applicant) that the applicant has met all responsibilities under the Emergency Planning and Community Right To Know Act of 1986, Title III, Public Law 99-499, in accordance with 10 CFR 70.22(i)(3)(xiii)

8.4.3.1.2 Onsite and Offsite Emergency Facilities

The emergency plan should list and describe onsite and offsite facilities that could be relied on in an emergency. The emergency plan should include the following:

---

[1] The map should include the location of sensitive facilities near the site, such as hospitals, schools, nursing homes, nearest residents, fire department, prisons, environmental sampling locations, and other structures and facilities that are important to emergency management.

- a list and description of both onsite and offsite emergency facilities, by location and purpose

- a description of emergency monitoring equipment that is available for personnel and area monitoring and for assessing the release to the environment of radioactive or hazardous chemicals incident to the processing of licensed material

- a description of the onsite and offsite services that support emergency response operations, including the following:

  - decontamination facilities
  - medical treatment facilities
  - first aid personnel
  - fire fighters
  - law enforcement assistance
  - ambulance services

- the applicant's commitment to the following:

  - facilities of adequate size and appropriate location that are designated, equipped, and ready for emergency use

  - adequate backup facilities required by the emergency plan and supporting documents that are available and ready for use

  - appropriate equipment and supplies necessary to support emergency response activities that are accessible during accident conditions

  - emergency equipment that is inventoried, tested, and serviced on a periodic basis to ensure accountability and reliability

  - sufficient reliable primary and backup communications channels available to accommodate emergency needs

  - offsite emergency resources and services that are identified and ready to ensure their timely mobilization and use

  - operational engineering information, such as current as-built drawings and procedures, that are readily available in the emergency facilities

  - sufficient equipment for personnel protection and monitoring

  - systems in place to alert onsite and offsite personnel in case of an emergency

8.4.3.1.3  Types of Accidents

For each general type of accident identified in the ISA Summary for which protective actions may be needed, the emergency plan should describe the following:

- the process and physical location(s) where the accidents could occur

- complicating factors and possible onsite and offsite consequences, including releases of nonradioactive hazardous chemicals incident to the processing of licensed material that could impact emergency response efforts

- the accident sequence that has the potential for the greatest radiological or toxic chemical impact

- figure(s) projecting doses and toxic substance concentrations as a function of distance and time for various meteorological stability classes, including a description of how the applicant projected such doses or concentrations (e.g., computer models and assumptions)

8.4.3.1.4 Classification of Accidents

The emergency plan should classify accidents as follows:

- The emergency plan classification system should include the following two classifications:

  (1) Alert: Events that may occur, are in progress, or have occurred that could lead to a release of radioactive material or hazardous chemicals incident to the processing of licensed material; however, the release is not expected to require a response by an offsite response organization to protect persons off site.

  (2) Site area emergency: Events that may occur, are in progress, or have occurred that could lead to a significant release of radioactive material or hazardous chemicals incident to the processing of licensed material and that could require a response by offsite emergency response organizations to protect persons off site.

- The emergency plan should identify the classification (alert or site area emergency) expected for each accident identified in the emergency plan.

- The emergency plan should specify emergency action levels (EALs) at which an alert or site area emergency will be declared. EALs are specific conditions that require the performance of emergency response measures. The applicant's EALs should be consistent with Appendix A to Regulatory Guide 3.67 and should be comparable to the U.S. Environmental Protection Agency (EPA) Protective Action Guides described in EPA 400-R-92-001, "Manual of Protective Action Guides and Protective Actions for Nuclear Incidents," issued May 1992. Transportation accidents more than 1.61 km (1 mi) from the facility should not be classified.

- The emergency plan should designate the personnel positions and alternates with the responsibility for accident classification during normal operations and back shifts.

8.4.3.1.5 Detection of Accidents

For each type of accident identified, the emergency plan should describe the following:

- the means of detecting the accident

- the means of detecting any release of radioactive material or hazardous chemicals incident to the processing of licensed material

- the means of alerting the operating staff

- the anticipated response of the operating staff

### 8.4.3.1.6 Mitigation of Consequences

For each accident identified in the ISA Summary, the emergency plan should briefly describe measures and equipment to be used for safe shutdown and the mitigation of consequences to workers on site and off site and to the public off site.

### 8.4.3.1.7 Assessment of Releases

The emergency plan should describe the following aspects of the applicant's procedures to be used to promptly and effectively assess the release of radioactive material or hazardous chemicals incident to the processing of licensed material:

- procedures for estimating or measuring the release rate or source term

- valid computer codes used to project doses or concentrations to the public or environment and their associated assumptions, along with adequate justifications to show the validity of the assumptions

- types, methods, frequencies, implementation times, and other details of onsite and offsite sampling and monitoring that will be performed to assess a release of radioactive materials or hazardous chemicals incident to the processing of licensed material

- the method for assessing collateral damage to the facility (especially items relied on for safety)

The emergency plan should describe the applicant's procedure for validating any code used to assess releases of radioactive material or hazardous chemicals incident to the processing of licensed material.

### 8.4.3.1.8 Responsibilities

The emergency plan should describe the emergency response organization and administration that ensure effective planning, implementation, and control of emergency preparedness activities. In addition, the applicant should make the following commitments:

- Procedures will clearly define the organizational structure and chain of command.

- Staffing and resources will be sufficient to accomplish all assigned tasks.

- Procedures will clearly define responsibilities and authority for each management, supervisory, and professional position. Responsibility is assigned for the coordination of onsite and offsite emergency response preparedness.

- Procedures will clearly define interfaces with supporting groups, both on site and off site.

- Mutual cooperation agreements exist or will be entered into with local agencies, such as fire, police, ambulance and rescue, and medical units.

- Plant management measures will be in place through procedures to audit and assess emergency preparedness to ensure site readiness to handle emergencies and to identify and correct problems.

- The onsite emergency response organization will provide effective command and control of the site during the assessment, mitigation, and recovery phase of an accident.

- The emergency public information system will provide advance and ongoing information to the media and public on subjects that would be discussed during an emergency, such as radiation hazards, chemical hazards, site operation, and site emergency plans.

- The schedule of emergency preparedness procedure development will ensure that procedures are available to support startup and operation of new processes and facilities on site.

8.4.3.1.9 Notification and Coordination

The emergency plan should provide reasonable assurance that emergency notification procedures will enable the emergency organization to correctly classify emergencies, notify emergency response personnel, and initiate or recommend appropriate actions in a timely manner, on the basis of the following:

- Emergency events are classified on the basis of the current emergency plan.

- Notification procedures minimize distraction of shift operating personnel and include concise, preformatted messages. Appropriate followup messages to offsite authorities are issued promptly.

- Information on the nature and magnitude of the hazards is made available to the appropriate emergency response personnel.

- Radiological and chemical source term data are available to the command post, technical support center, emergency operation center, and appropriate State personnel in cooperation with the NRC.

- When available, offsite field monitoring data are logged, compared with source term data, and used in the protective action recommendation process.

- Protective Action Guides are available and are used by the appropriate personnel in a timely manner.

- The emergency public information program ensures timely dissemination of accurate, reliable, and understandable information.

- Systems are in place, if required, to alert, notify, and mobilize onsite and offsite response personnel in case of an emergency.

- Procedures are in place to notify and coordinate with responsible parties when some personnel, equipment, and facility components are not available.

The emergency plan should describe who will take the following actions and how he or she will act promptly and effectively:

- the decision to declare an alert or site area emergency

- the activation of the onsite emergency response organization during all shifts

- the prompt notification of offsite response authorities that an alert or site area emergency has been declared, including the licensee's initial recommendation for offsite protective actions (normally within 15 minutes of classification)

- the notification to the NRC Operations Center (as soon as possible and, in any case, no later than 1 hour after a declared emergency)

- the decision regarding which onsite protective actions to initiate

- the decision regarding which offsite protective actions to recommend

- the decision to request support from offsite organizations

- the decision to terminate the emergency or enter recovery mode

8.4.3.1.10 Information To Be Communicated

The emergency plan should describe the information to be communicated during an emergency, including the following:

- a standard reporting checklist to facilitate timely notification

- the types of information to be provided concerning facility status, radioactive releases or hazardous chemicals incident to the processing of licensed material, and protective action recommendations

- a description of preplanned protective action recommendations to be made to each appropriate offsite organization

- the offsite officials to be notified, as a function of the classification of the event

- the recommended actions to be taken by offsite organizations for each accident treated in the emergency plan

### 8.4.3.1.11 Training

The emergency plan should describe the frequency, performance objectives, and plans for the emergency response training that the licensee will provide to workers. The plan should include the following:

- the topics and general content of training programs for the licensee's onsite and offsite emergency response personnel to satisfy the objectives described above

- the administration of the training program, including responsibility for training, the positions to be trained, the schedules for training, the frequency of retraining, the use of team training, and the estimated number of hours of initial training and retraining

- the training to be provided on the use of protective equipment, such as respirators, protective clothing, monitoring devices, and other equipment used in emergency response

- the training program for onsite personnel who are not members of the emergency response staff

- any special instructions and orientation tours that the licensee would offer to fire, police, medical, and other nonlicensee emergency personnel who may be required to respond to an emergency to ensure that they know the emergency plan, assigned duties, and effective response to an actual emergency

### 8.4.3.1.12 Safe Shutdown (Recovery and Facility Restoration)

The emergency plan should describe the following aspects of the applicant's plans for adequately restoring the facility to a safe status after an accident and recovery after an emergency:

- the methods and responsibilities for assessing the damage to and status of the facility's capabilities to safely control radioactive material or hazardous chemicals associated with the process

- the procedures for promptly determining the actions necessary to reduce any ongoing releases of radioactive material or hazardous chemicals incident to the processing of licensed material and to prevent further incidents

- the provisions for promptly and effectively accomplishing required restoration actions

- key positions in the recovery organization

### 8.4.3.1.13 Exercises and Drills

The emergency plan should state the applicant's commitment to conduct exercises and drills in a manner that demonstrates the capability of the organization to plan and perform an effective response to an emergency. An adequate plan should demonstrate the following:

- Qualified individuals for each position in the emergency response organization demonstrate task-related knowledge through periodic participation.

- Drill performance is assessed against specific scenario objectives using postulated accidents that adequately test personnel, equipment, and resources, including previously identified weaknesses.

- Effective player, controller, evaluator, and observer predrill briefings are conducted.

- Scenario data and exercise messages provided by the controllers effectively maintain the timeline and do not interfere with the emergency organization's response to exercise scenario events, except where safety considerations are involved.

- Trained evaluators are used to identify and record participant performance, scenario strengths and deficiencies, and equipment problems.

- The prestaging of equipment and personnel is minimized to realistically test the activation and staffing of emergency facilities.

- Critiques are conducted promptly and include a followup plan for correcting any identified weaknesses and improving training effectiveness.

- Emergency drills demonstrate that resources are effectively used to control the site, mitigate further damage, control radiological releases, perform required onsite activities under simulated radiation/airborne and other emergency conditions, accurately assess the facility's status during an accident, and initiate recovery.

- Emergency drills demonstrate personnel protection measures, including controlling and minimizing hazards to individuals during fires, medical emergencies, mitigation activities, search and rescue, and other similar events.

- The emergency drills demonstrate that onsite communications effectively support emergency response activities.

- The emergency drills demonstrate that the emergency public information organization disseminates accurate, reliable, timely, and understandable information.

- Provisions are made for conducting quarterly communications checks with offsite response organizations.

- Offsite organizations are invited to participate in the biennial onsite exercise, which tests the major elements of the emergency plan and response organizations.

8.4.3.1.14 Responsibilities for Developing and Maintaining the Emergency Program and Its Procedures

The emergency plan should describe the following aspects of the responsibilities for developing and maintaining the emergency program and its procedures:

- the means for ensuring that revisions to the emergency plan and the procedures used to implement the emergency plan are adequately prepared, kept up to date (normally within 30 days of any changes), and distributed to all affected parties, including the NRC

- the provisions for approving the implementing emergency procedures, making and distributing changes to the procedures, and ensuring that each person responsible for an emergency response function has immediate access to a current copy of the emergency procedures[2]

- procedures for allowing offsite response organizations 60 days to comment on any new emergency plan or significantly updated emergency plans[3]

### 8.4.3.2 Evaluation That No Emergency Plan Is Required

The staff should review the adequacy of the evaluation (or the referenced ISA Summary) that no emergency plan is required against the requirements in 10 CFR 70.22(i)(2) and the specific criteria provided in Sections 8.4.3.2.1 through 8.4.3.2.4 of this SRP. This evaluation should be acceptable if it meets the regulatory requirements and the acceptance criteria given below.

### 8.4.3.2.1 Facility Description

The evaluation should describe the facility and site, the area near the site, and the licensed activities conducted at the facility. To be considered sufficient to support the evaluation, these descriptions should include the following:

- a detailed drawing of the site showing (1) onsite and near offsite (within 1.61 km (1 mi)) structures, with building numbers and labels, (2) roads and parking lots on site and main roads near the site, (3) site boundaries, including fences and gates, (4) major site features, (5) water bodies within approximately 1.61 km (1 mi), and (6) the location(s) of the nearest residents

- the stack heights, typical stack flow rates, and efficiencies of any emission control devices

- a general description of licensed and other major activities conducted at the facility and the type, form, and quantities of radioactive material used

### 8.4.3.2.2 Types of Accidents

The evaluation should describe or refer to each type of accident identified by the ISA Summary that has maximum offsite consequences that exceed the limit specified in 10 CFR 70.22(i)(1)(i). In addition, the following information should be available for review:

- the process and physical location where the accident could occur

---

[2] Provisions for approving changes to the emergency plan and the procedures and the individuals authorized to make those changes should be clearly stated.

[3] The applicant does not need to provide offsite organizations with amendments to emergency plans that do not affect an organization and those allowed by 10 CFR 70.32(i) before it submits them to the NRC.

- complicating factors and offsite consequences, including the release of nonradioactive hazardous chemicals incident to the processing of licensed material

- the accident sequence that has the potential for the greatest radiological and toxic chemical impact

### 8.4.3.2.3 Detection of Accidents

For each type of accident identified, the evaluation should describe the following:

- the means of detecting the accident

- the means of detecting any release of radioactive or hazardous chemicals incident to the processing of licensed material

- the means of alerting the operating staff

- the anticipated response of the operating staff

### 8.4.3.2.4 Evaluation of Maximum Public Exposure

To demonstrate that no emergency plan is required, an applicant may either (1) request that its total possession limit for radioactive material be reduced below the emergency plan threshold in 10 CFR 70.22(i)(1) or (2) perform a site-specific evaluation (or refer to the ISA Summary, as appropriate) to demonstrate that maximum public exposure is less than the limits specified in 10 CFR 70.22(i)(1)(i).

The evaluation should make available the following information sufficient to allow for independent verification:

- the type of accident (e.g., fire, explosion, hazardous chemicals released that are incident to the processing of licensed material, and nuclear criticality)

- the location of the accident

- the maximum source term

- the solubility of material

- the facility design or items relied on for safety and the proposed release fraction

- the location and distance of the nearest member of the public to the facility

- the dose model and process used to verify the reliability of the model and the validity of the assumptions

- the assumed worst case weather condition

- the maximum calculated exposure to a member of the public at the facility boundary

The evaluation should list and describe the factors in 10 CFR 70.22(i)(2) that the applicant considered in evaluating the maximum dose to members of the public. The applicant should demonstrate why the factors used in the evaluation are appropriate when compared with the factors in NUREG-1140. If the factors and evaluation show that the maximum dose to a member of the public off site from a release of radioactive materials could not exceed 0.01-Sv (1-rem) effective dose equivalent or the intake of 2 mg of soluble uranium, no emergency plan is required in accordance with 10 CFR 70.22(i)(1)(i).

### 8.4.3.3 Requirements for New Facilities or New Processes at Existing Facilities

The application should address the BDC for new facilities or new processes at existing facilities that require a license amendment under 10 CFR 70.72, "Facility Changes and Change Process." Regulations in 10 CFR 70.64 (a)(6) require that the BDC must be applied to the design of new processes but do not require retrofits to existing facilities or existing processes; however, all facilities and processes must comply with the performance requirements in 10 CFR 70.61, "Performance Requirements." The licensee or applicant should clearly state how the design of the facility or process provides for the emergency capability to maintain control of the following:

- licensed material and hazardous chemicals produced from licensed material, including:

    - how the site, facility, or process was designed to process and store both licensed materials and hazardous chemicals either produced from licensed material or used in the process (referencing the licensee or applicant's sitewide security plan, if desired)

    - controls on the interface of hazardous chemicals with the licensed material process safety program

- evacuation of onsite personnel, including the criteria used designing the facility to allow personnel to evacuate (e.g., time, dose, ease of egress)

- onsite emergency facilities and services that facilitate the use of available offsite services, including:

    - the offsite services that will be needed in an emergency at the facility

    - the criteria used to design the facility to detect accidents

    - the criteria used to design the facility to alert facility staff of an accident

    - the criteria was used to design the facility to notify and coordinate with both onsite and offsite personnel

    - the criteria used to design the facility to allow for the transportation of personnel to onsite and offsite facilities or locations

## 8.5 Amendments or Changes to the Emergency Plan

The applicant may make changes to the approved emergency plan without NRC approval if the changes do not decrease the effectiveness of the plan and the applicant submits copies of the changes to the NRC and appropriate organizations within 6 months of making the changes in accordance with 10 CFR 70.32(i).  The applicant may not implement proposed changes that decrease the effectiveness of the emergency plan without prior application to and approval of the NRC.

## 8.6 Review Procedures

### 8.6.1 Acceptance Review

The primary reviewer should evaluate the application to determine whether it addresses the areas of review discussed in Section 8.3 above.  If significant deficiencies are identified, the reviewer(s) should ask the applicant to submit additional material before the safety evaluation begins.

### 8.6.2 Safety Evaluation

After determining that the application is acceptable for review in accordance with Section 8.6.1 above, the primary reviewer should perform a safety evaluation against the acceptance criteria described in Section 8.4 above.  If the primary reviewer identifies the need for additional information during the safety evaluation, the primary reviewer should coordinate with the licensing project manager in preparing a request for additional information.

#### 8.6.2.1 Emergency Plan

After the NRC staff receives an acceptable application from the applicant, the primary reviewer should conduct a complete review of the applicant's emergency plan and assess its acceptability in accordance with Section 8.4.3.1 above.  The reviewer should verify that emergency planning is consistent with the potential accident sequences described in the ISA Summary.  The ISA Summary reviewer and emergency plan reviewer should coordinate their efforts to ensure the resolution of any issues concerning the emergency plan relative to ISA Summary information.

Although the section of the licensee's submittal entitled "Emergency Management Program" should contain the bulk of this information, the primary and secondary reviewers should gain familiarity with the site, including its demography, land use, facility design and layout, and major accidents postulated by the applicant, as presented in relevant sections of the application.  The primary and secondary reviewers should also become familiar with proposed radiation protection activities and other operational matters that interface with emergency plans (particularly the functions reviewed using SRP Chapters 4 and 11).  The reviewers should consult draft and final environmental statements for the proposed facility.  This information may be supplemented by a personal visit to the site by the reviewer and meetings with the applicant.  As the final step, the primary reviewer should prepare a safety evaluation report (SER) section in accordance with Section 8.7 below.

*8.6.2.2 Evaluation That No Emergency Plan Is Required*

The primary reviewer should verify that the evaluation is consistent with the potential accident sequences described in the ISA Summary. The ISA Summary reviewer and the primary reviewer should coordinate their efforts to ensure the resolution of any issues concerning the evaluation relative to ISA information. As the final step, the primary reviewer should prepare an SER section in accordance with SRP Section 8.7 that either agrees with the applicant's conclusion that no emergency plan is required or indicates that the staff does not accept the applicant's evaluation and recommends that an emergency plan be required.

## 8.7  Evaluation Findings

The primary reviewer writes an SER section addressing each topic reviewed under this SRP chapter and explains why the NRC staff has reasonable assurance that the emergency management part of the application is acceptable. The agency may propose license conditions to impose requirements where the application is deficient. The report includes a summary statement describing what was evaluated and why the reviewer finds the submittal acceptable. The staff can document the evaluation as follows:

> The staff has evaluated [insert a summary statement describing what was evaluated and why the reviewer finds the submittal acceptable]. In accordance with 10 CFR 70.22(i), the licensee commits to maintain and execute an emergency plan for responding to the radiological hazards resulting from a release of radioactive material or hazardous chemicals incident to the processing of licensed material. The NRC staff reviewed the emergency plan with respect to 10 CFR 70.22(i) and the acceptance criteria in SRP Section 8.4.3. The NRC staff determined that the applicant's emergency plan is adequate to demonstrate compliance with 10 CFR 70.22(i) in that (1) the facility is properly configured to limit releases of radioactive materials in the event of an accident, (2) a capability exists for measuring and assessing the significance of accidental releases of radioactive materials, (3) appropriate emergency equipment and procedures are provided on site to protect workers against radiation and other chemical hazards that might be encountered after an accident, (4) a system has been established to notify Federal, State, and local government agencies and to recommend appropriate protective actions to protect members of the public, and (5) the necessary recovery actions are established to return the facility to a safe condition after an accident.

> The requirements of the emergency plan are implemented through approved written procedures. Changes that decrease the effectiveness of the emergency plan may not be made without NRC approval. The NRC will be notified of other changes that do not decrease the effectiveness of the emergency plan within 6 months of making the changes.

## 8.8  References

*U.S. Code of Federal Regulations*, Chapter I, Title 10, "Energy," Part 70, "Domestic Licensing of Special Nuclear Material."

U.S. Nuclear Regulatory Commission, "Part 30 Statements of Consideration and Emergency Preparedness for Fuel Cycle and Other Radioactive Material Licensees," *Federal Register*, 54 FR 14051 (April 7, 1989.)

U.S. Nuclear Regulatory Commission, "Nuclear Fuel Cycle Accident Analysis Handbook," NUREG/CR-6410, March 1998.

U.S. Nuclear Regulatory Commission, "Response Technical Manual (RTM) 96," NUREG/BR-0150, Vol. 1, Revision 4, 1996.

U.S. Environmental Protection Agency, "Manual of Protective Action Guides and Protective Actions for Nuclear Incidents," EPA 400-R-92-001, May 1992.

# 9. ENVIRONMENTAL PROTECTION

## 9.1 Purpose of Review

The purpose of this review is to determine whether the applicant's proposed environmental protection measures are adequate to protect the environment and public health and safety and to comply with the regulatory requirements imposed by the U.S. Nuclear Regulatory Commission (NRC) in Title 10 of the *Code of Federal Regulations* (10 CFR) Part 20, "Standards for Protection against Radiation" and 10 CFR Part 70, "Domestic Licensing of Special Nuclear Material."

This chapter does not address the specific requirements of 10 CFR Part 51, "Environmental Protection Regulations for Domestic Licensing and Related Regulatory Functions." NUREG-1748, "Environmental Review Guidance for Licensing Actions Associated with NMSS Programs," issued August 2003, provides general procedures for the environmental review of licensing actions regulated by the Office of Nuclear Material Safety and Safeguards (NMSS). The staff of the Division of Fuel Cycle Safety and Safeguards (FCSS) should coordinate the preparation of an environmental assessment (EA) and finding of no significant impact (FONSI) or an environmental impact statement (EIS) with the Office of Federal and State Materials and Environmental Management Programs. If the licensee proposes that a requested action is a categorical exclusion under the provisions of 10 CFR 51.22, "Criterion for Categorical Exclusion; Identification of Licensing and Regulatory Actions Eligible for Categorical Exclusion or Otherwise Not Requiring Environmental Review," the FCSS staff should confirm that the action meets the applicable criteria in 10 CFR 51.22(c).

## 9.2 Responsibility for Review

Primary: Environmental Engineer/Scientist

Secondary: Licensing Project Manager

Supporting: Fuel Cycle Facility Inspector
Radiation Safety Reviewer
Integrated Safety Analysis (ISA) Primary Reviewer

## 9.3 Areas of Review

The environmental safety program should address the environmental protection measures, including the control and monitoring of gaseous and liquid effluents and the management of solid waste. The environmental program should also provide for the monitoring of the facility environment, including ambient air, surface water, ground water, soils, and vegetation that can be affected by facility effluents. This chapter addresses the areas of review for environmental protection measures, and for environmental monitoring measures.

If the application includes an ISA Summary as required by Subpart H, "Additional Requirements for Certain Licensees Authorized To Possess a Critical Mass of Special Nuclear Material," of 10 CFR Part 70, the environmental reviewer will review the ISA Summary accident sequences that could result in high or intermediate consequences to an individual located outside the controlled area or that could result in a 24-hour averaged release of radioactive material outside the restricted area in concentrations exceeding 5,000 times the values in Table 2 of Appendix B,

"Annual Limits on Intake (ALIs) and Derived Air Concentrations (DACs) of Radionuclides for Occupational Exposure; Effluent Concentrations; Concentrations for Release to Sewerage," to 10 CFR Part 20. Section 9.3.2.3 below addresses areas of review for the ISA Summary specific to environmental protection.

The regulatory requirements for environmental protection appear in 10 CFR Part 20, 10 CFR Part 51, and 10 CFR Part 70. The NRC staff focuses its environmental review on that part of the plant wide safety program that the applicant establishes to control and assess the level of radioactive and nonradioactive releases (gaseous, liquid, and solid) to the environment. Therefore, the staff reviews the effluent control portion of the applicant's radiation protection program and the applicant's effluent and environmental monitoring practices.

To receive authorization to possess a critical mass quantity of special nuclear material (SNM), an applicant must also perform an ISA and prepare an ISA Summary in accordance with Subpart H of 10 CFR Part 70. SRP Chapter 3 presents guidance on the ISA. The environmental safety review of the ISA Summary will examine the identified potential accident sequences that result in radiological and non-radiological releases to the environment, the items relied for safety (IROFS) that the applicant specifies to reduce the risk of those accidents, and the associated management measures that provide reasonable assurance that the IROFS will perform their designated safety functions.

Thus, environmental protection encompasses three main components, as necessary: (1) effluent and environmental controls and monitoring, (2) the ISA Summary and other ISA documentation as described in Sections 9.3.1 through 9.3.2 below, and (3) management measures in the license application.

### 9.3.1 Effluent and Environmental Controls and Monitoring

The staff's review of the environmental radiation protection program described in the application encompasses the following areas:

- as low as reasonably achievable (ALARA) goals for effluent control

- effluent controls to maintain public doses ALARA

- ALARA reviews and reports to management

- waste minimization practices and, for new operations, design plans for waste minimization

The staff's review of the applicant's effluent and environmental monitoring practices described in the application encompasses the following areas:

- in-place filter-testing procedures for air-cleaning systems

- known or expected concentrations of radionuclides in effluents

- physical and chemical characteristics of radionuclides in discharges

- discharge locations

- environmental media to be monitored and the sample locations

- sampling collection and analysis procedures, including the minimum detectable concentrations of radionuclides

- action levels and actions to be taken when the levels are exceeded

- permits, including air discharge and National Pollutant Discharge Elimination System permits

- leak detection systems for ponds, lagoons, and tanks

- pathways analysis methods to estimate public doses

- recording and reporting procedures

- solid waste handling and disposal programs

### 9.3.2 Integrated Safety Analysis Summary

The staff's review of the applicant's ISA Summary related to environmental protection includes the following areas:

- accident sequences (and associated facility processes) that, if unmitigated, would result in releases to the environment

- likelihood and environmental consequences of these accident sequences

- controls relied on to reduce the unmitigated risk from high or intermediate risk to an acceptable level

- availability and reliability of controls

### 9.3.3 Environmental Protection Management Measures

The staff's review of the applicant's management measures related to environmental protection includes the following areas:

- a method for grading management measures commensurate with the reduction in risk attributable to each control or control system

- a commitment to design, implement, and maintain the controls and control systems to ensure that they are available and reliable to perform their functions when needed

Review Interfaces

In addition to Chapter 9 of the application, the environmental reviewer should examine information in the following other areas to ensure that it is consistent with the information in Chapter 9:

- facility and process descriptions applied to environmental protection as described in SRP Chapter 1

- the safety program, ISA commitments, and ISA documentation applied to environmental protection as described in SRP Chapter 3

- the radiation safety program as described in SRP Chapter 4

- chemical processes applied to environmental protection as described in SRP Chapter 6

- fire-initiated accident sequences that have the potential to result in high or intermediate consequences as described in SRP Chapter 7

- configuration management, maintenance, training and qualifications, procedures, audits and assessments, incident investigations, record management, and other quality assurance elements as described in SRP Chapter 11[1]

## 9.4  Acceptance Criteria

Sections 9.4.3.1 through 9.4.3.3 describe acceptance criteria for the effluent and environmental controls and monitoring, the ISA Summary, and management measures.

### 9.4.1 Regulatory Requirements

To be considered acceptable, the application must satisfy the following regulatory requirements for environmental protection:

- Subpart B, "Radiation Protection Programs"; Subpart D, "Radiation Dose Limits for Individual Members of the Public"; and Subpart F, "Surveys and Monitoring," of 10 CFR Part 20 specify the effluent control and treatment measures necessary to meet the dose limits and dose constraints for members of the public. Subpart F also states the survey requirements. Subpart K, "Waste Disposal," specifies the waste disposal requirements; Subpart L, "Records," specifies the records requirements; and Subpart M, "Reports," specifies the reporting requirements.

- 10 CFR Part 51 provides that the applicant must establish effluent and environmental monitoring systems to provide the information required by 10 CFR 51.60(a).

- 10 CFR Part 70 requires the applicant to demonstrate that proposed facilities and equipment, including measuring and monitoring instruments and devices for the disposal of radioactive effluents and wastes, are adequate to protect the environment and public health and safety, as specified in 10 CFR 70.22(a)(7).

---

[1]   Section 9.3.3 addresses areas of review for management measures applied to environmental protection.

- 10 CFR Part 70 also provides that the applicant for a facility (as described in 10 CFR 70.4, "Definitions") must submit a safety assessment of the design basis of the principal structures, systems, and components of the plant, including provisions for protection against natural phenomena, as specified in 10 CFR 70.22(f).

- 10 CFR Part 70 also provides that an applicant for a facility must provide an ISA Summary that includes a list of the IROFS established by the applicant and other elements, as described in 10 CFR 70.65(b).

- 10 CFR 70.59, "Effluent Monitoring Reporting Requirements," outlines the reporting requirements for radiological effluent monitoring for a 10 CFR Part 70 licensee.

## 9.4.2 Regulatory Guidance

The regulatory guidance for environmental protection appears in the following NRC and industry documents:

- NRC Regulatory Guide 4.15, "Quality Assurance for Radionuclide Monitoring Programs (Inception through Normal Operations to License Termination)—Effluent Streams and the Environment," Revision 2, July 2007

- NRC Regulatory Guide 4.16, "Monitoring and Reporting Radioactivity in Releases of Radioactive Materials in Liquid and Gaseous Effluents from Nuclear Fuel Processing and Fabrication Plants and Uranium Hexafluoride Production Plants," December 1985

- NRC Regulatory Guide 4.20, "Constraint on Releases of Airborne Radioactive Materials to the Environment for Licensees Other Than Power Reactors," December 1996

- NRC Regulatory Guide 8.37, "ALARA Levels for Effluents from Materials Facilities," July 1993

- NRC Information Notice 94-07, "Solubility Criteria for Liquid Effluent Releases to Sanitary Sewerage under the Revised 10 CFR Part 20," January 28, 1994

- NRC Information Notice 94-23, "Guidance to Hazardous, Radioactive, and Mixed Waste Generators on the Elements of a Waste Minimization Program," March 25, 1994

- American National Standards Institute (ANSI) N13.1-1982, "Guide to Sampling Airborne Radioactive Materials in Nuclear Facilities"

- ANSI N42.18-1980, "Specification and Performance of On-Site Instrumentation for Continuously Monitoring Radioactive Effluents"

- National Council on Radiation Protection and Measurements (NCRP) Report No. 123, "Screening Models for Releases of Radionuclides to Atmosphere, Surface Water, and Ground," Volumes I and II, January 1996

- NUREG/CR-6410, "Nuclear Fuel Cycle Accident Analysis Handbook," March 1998

- NUREG-1513, "Integrated Safety Analysis Guidance Document," May 2001

- NUREG-1748, "Environmental Review Guidance for Licensing Actions Associated with NMSS Programs," August 2003

### 9.4.3 Regulatory Acceptance Criteria

*9.4.3.1 Environmental Report or Categorical Exclusion*

An environmental report is required for actions listed in 10 CFR 51.60(b). NUREG-1748 discusses the acceptance criteria for the environmental report.

An environmental report is not required for licensing actions that meet the requirements for a categorical exclusion, as defined in 10 CFR 51.22(c). However, if, under 10 CFR 51.23(c)(11), the action involves an amendment to licenses for fuel cycle plants, radioactive waste disposal sites, and other materials licenses identified in 10 CFR 51.60(b)(1) for changes in process operations or equipment, the applicant must demonstrate that the action will not result in significant effects on the environment. NUREG-1748 gives the acceptance criteria for this categorical exclusion.

If a license application indicates a significant increase in the potential for, or consequences of, radiological accidents, then the licensing action is NOT categorically excluded from review under the National Environmental Policy Act. The application must include an environmental report, and the staff must prepare an EA.

*9.4.3.2 Effluent and Environmental Controls and Monitoring*

An applicant's proposed environmental protection measures are acceptable if they provide for qualified and trained staff, effluent control, and effluent and environmental monitoring in accordance with the NRC's requirements. Using the acceptance criteria defined in SRP Chapter 11, the NRC staff will review qualifications and training that the applicant has established for plant personnel who are associated with environmental protection. This review will include the qualification and training of managers, supervisors, technical staff, operators, technicians, and maintenance personnel whose levels of knowledge are important to the environment and protect public health and safety. The NRC will expect managers and staff to have levels of education and experience commensurate with the responsibilities of their positions.

9.4.3.2.1 Effluent Controls and Waste Minimization

In accordance with 10 CFR 20.1101, "Radiation Protection Programs," each licensee must implement a radiation protection program, which is discussed in detail in SRP Chapter 4. The environmental review of the radiation protection program focuses on the applicant's methods to maintain *public* doses ALARA in accordance with 10 CFR 20.1101. NRC guidance on compliance with these regulations appears in Regulatory Guide 8.37.

Specifically, 10 CFR 20.1101(d) requires the applicant to establish constraints on airborne emissions of radioactive material to the environment, excluding radon-222 and its decay products. Such constraints must ensure that the individual member of the public who is likely to receive the highest dose will not be expected to receive a total effective dose equivalent (TEDE) in excess of 0.1 millisievert (10 millirem) per year from these emissions. To meet the reporting requirements of 10 CFR 20.2203, "Reports of Exposures, Radiation Levels, and Concentrations of Radioactive Material Exceeding the Constraints or Limits," the applicant must have (and describe) procedures for reporting to the NRC when these dose constraints are exceeded and must take prompt appropriate corrective action to prevent recurrence. NRC guidance on compliance with this regulation can be found in Regulatory Guide 4.20.

The environmental review of the radiation protection program also focuses on the applicant's waste minimization practices. Applicants for new licenses are required to comply with 10 CFR 20.1406, "Minimization of Contamination," which states that the applicant must describe how facility design procedures for operation will, to the extent practicable, minimize contamination of the facility and the environment, facilitate eventual decommissioning, and minimize the generation of radioactive waste. Applicants requesting amendment or renewal of existing licenses must minimize and control waste generation during operations as part of the radiation protection program, in accordance with 10 CFR 20.1101 (Volume 62 of the *Federal Register*, page 39,082 (62 FR 39082); July 21, 1997).

NRC Information Notice 94-23 offers guidance for waste minimization programs. SRP Chapter 10 offers more information on compliance with the decommissioning aspects of the waste minimization regulations.

The proposed radiation protection program is acceptable if it satisfies the following criteria:

- Radiological (ALARA) Goals for Effluent Control

    ALARA goals are set at a modest fraction (10 to 20 percent) of the values in Table 2, Columns 1 and 2, and Table 3 of Appendix B to 10 CFR Part 20 and the external exposure limit in 10 CFR 20.1302(b)(2)(ii), or the dose limit for members of the public, if the applicant proposes to demonstrate compliance with 10 CFR 20.1301, "Dose Limits for Individual Members of the Public," through a calculation of the TEDE to the individual likely to receive the highest dose.

    An applicant's constraint approach is acceptable if it is consistent with guidance found in Regulatory Guide 4.20 and if the applicant's description of the constraint approach provides sufficient detail to demonstrate specific application of the guidance to proposed routine and nonroutine operations, including anticipated events.

- Effluent Controls To Maintain Public Doses ALARA

    The applicant describes and commits to the use of effluent controls (e.g., procedures, engineering controls, and process controls) to maintain public doses ALARA. Common control practices include filtration, encapsulation, adsorption, containment, recycling, leakage reduction, and storage of materials for radioactive decay. Practices for large, diffuse sources (such as contaminated soils or surfaces) include covers, wetting during operations, and the application of stabilizers. The applicant must demonstrate a commitment to reduce unnecessary exposure to members of the public and releases to the environment.

    Engineering options that do not substantially reduce the collective dose and require unreasonable costs are not required. "Reasonableness" can be founded on qualitative or quantitative cost/benefit analyses. Quantitative analyses may use a value of 2,000 per person-rem (person-centisievert), as discussed in NUREG-1530,

"Reassessment of the NRC's Dollar per Person-Rem Conversion Factor Policy," issued December 1995.

- ALARA Reviews and Reports to Management

  The applicant commits to an annual review of the content and implementation of the radiation protection program, which includes the ALARA effluent control program. This review includes analysis of trends in release concentrations, environmental monitoring data, and radionuclide usage; determines whether operational changes are needed to achieve the ALARA effluent goals; and evaluates all designs for system installations or modifications. The applicant also commits to reporting the results to senior management, along with recommendations for changes in facilities or procedures that are necessary to achieve ALARA goals.

- Waste Minimization

  To comply with 10 CFR 20.1406, applications for new licenses must describe how the facility's design procedures for operation will minimize, to the extent practicable, the contamination of the facility and the environment and the generation of radioactive waste. Waste minimization programs proposed by applicants for both new and existing licenses are acceptable if the programs include the following:

  - top management support

  - the methods used to characterize waste generation (including types and amounts) and waste management costs (including costs of regulatory compliance, paperwork, transportation, treatment, storage, and disposal)

  - periodic waste minimization assessments to identify waste minimization opportunities and solicit employee or external recommendations

  - provisions for technology transfer to seek and exchange technical information on waste minimization

  - the methods used to implement and evaluate waste minimization recommendations

### 9.4.3.2.2 Effluent and Environmental Monitoring

The applicant's effluent monitoring is considered acceptable if it meets the following criteria:

- The known or expected concentrations of radioactive materials in airborne and liquid effluents are ALARA *and* are below the limits specified in Table 2 of Appendix B to 10 CFR Part 20 or the site-specific limits established in accordance with 10 CFR 20.1302(c).

  If, in accordance with 10 CFR 20.1302(c), the applicant proposes to adjust the effluent concentrations in Appendix B to 10 CFR Part 20 to account for the actual physical and chemical characteristics of the effluents, the applicant must provide information on aerosol size distributions, solubility, density, radioactive decay equilibrium, and chemical form. This information must be complete and accurate to justify the derivation and application of the alternative concentration limits for the radioactive materials.

- If the applicant proposes to demonstrate compliance with 10 CFR 20.1301 using a calculation of the TEDE to the individual who is likely to receive the highest dose in accordance with 10 CFR 20.1302(b)(1), it must support the calculation of the TEDE by pathway analyses with appropriate models, codes, and assumptions that accurately represent the facility, site, and the surrounding area. In addition, the assumptions must be reasonable, input data must be accurate, all applicable pathways must be considered, and the results must be interpreted correctly.

  NCRP Report No. 123 provides acceptable methods for calculating the dose from radioactive effluents. The use of computer codes is acceptable for pathway analyses if the applicant can demonstrate that any code it has used has undergone validation and verification to demonstrate the validity of estimates developed using the codes for established input sets. Dose conversion factors are acceptable for use in the pathway analyses if they are based on the methodology described in International Commission on Radiological Protection 30, "Limits for Intakes of Radionuclides by Workers," 1982, as reflected in the U.S. Environmental Protection Agency's Federal Guidance Report No. 11, "Limiting Values of Radionuclide Intake and Air Concentration and Dose Conversion Factors for Inhalation, Submersion, and Ingestion," issued September 1988. Such methods are acceptable for determining the dose to the maximally exposed individual during normal facility operations and anticipated events.

- The applicant identifies and monitors all liquid and airborne effluent discharge locations and identifies monitoring locations. For those effluent discharge points that have input from two or more contributing sources within the facility, sampling each contributing source is considered necessary for effective process and effluent control.

- The applicant continuously samples airborne effluents from all routine and nonroutine operations and from anticipated events associated with the plant, including effluents from areas that are not used for processing SNM, such as laboratories, experimental areas, storage areas, and fuel element assembly areas.

  Effluents are sampled unless the applicant has established (by periodic sampling or other means) that radioactivity in the effluent is insignificant and will remain so. In such cases, the effluent is sampled at least quarterly to confirm that its radioactivity is not significant. For the purposes of this SRP, radioactivity in an effluent is significant if the concentration averaged over a calendar quarter is equal to 10 percent or more of the appropriate concentration listed in Table 2 of Appendix B to 10 CFR Part 20.

- The sample collection and analysis methods and frequencies are appropriate for the effluent medium and the radionuclide(s) being sampled. Sampling methods ensure that the applicant obtains representative samples using appropriate sampling equipment and sample collection and storage procedures. For liquid effluents, the applicant collects representative samples at each release point to determine the concentrations and quantities of radionuclides that are released to an unrestricted area, including discharges to sewage systems. For continuous releases, the applicant collects samples continuously at each release point. For batch releases, the applicant collects a representative sample of each batch. If the applicant uses periodic sampling in lieu of continual sampling, it shows that the samples are representative of actual releases. Monitoring instruments are calibrated at least annually, or more frequently if suggested by the manufacturer.

- The applicant performs radionuclide-specific analyses on selected composite samples unless either of the following criteria exists:

  – The gross alpha and beta activities are so low that individual radionuclides could not be present in concentrations greater than 10 percent of the concentrations specified in Tables 2 or 3 of Appendix B to 10 CFR Part 20.

  – The radionuclide composition of the sample is known through operational data, such as the composition of the feed material.

  Monitoring reports in which the quantities of individual radionuclides are estimated on the basis of methods other than direct measurement include an explanation and justification of how the results were obtained.

  Operational data may not be adequate for determining radionuclide concentration in certain cases. Such cases include, but are not limited to: (1) plants that process uranium in which extraction, ammonium diuranate precipitation, ion exchange, or other separation process could result in the concentration of thorium isotopes (principally thorium-234), (2) plants that process uranium of varying enrichments, and (3) plants that process plutonium in which significant variation in the plutonium-238/plutonium-239 ratio among batches and the continuous ingrowth of americium-241 would preclude the use of feed material data to determine the radionuclide composition of effluents.

  The applicant performs radionuclide analyses more frequently than usual (1) at the beginning of the monitoring program until it establishes a predictable and consistent radionuclide composition in effluents, (2) whenever there is a significant, unexplained increase in gross radioactivity in effluents, and (3) whenever a process change or other circumstance might cause a significant variation in the radionuclide composition.

- The minimum detectable concentration (MDC) for sample analyses is not more than 5 percent of the concentration limits listed in Table 2 of Appendix B to 10 CFR Part 20. If the actual concentrations of radionuclides in samples are known to be higher than 5 percent of the 10 CFR Part 20 limits, the analysis methods need only be adequate to measure the actual concentration. However, in such cases, the MDC must be low enough to accommodate fluctuations in the concentrations of the effluent and the uncertainty of the MDC.

- The laboratory quality control procedures are adequate to validate the analytical results. These procedures include the use of established standards, such as those provided by the National Institute of Standards and Technology, and standard analytical procedures, such as those established by the National Environmental Laboratory Accreditation Conference.

- The proposed action levels and actions to be taken if the action levels are exceeded are appropriate. The action levels are incremental, such that each increasing action level results in a more aggressive action to ensure effluent control. A slightly higher than normal concentration of a radionuclide in an effluent triggers an investigation into the cause of the increase. The specified action level will result in the shutdown of an operation if the specified level is exceeded. These action levels are selected on the basis of the likelihood that a measured increase in concentration could indicate potential violation of the effluent limits.

- The applicant completely and accurately describes all applicable Federal and State standards for discharges and any permits issued by Federal, State, or local governments for gaseous and liquid effluents.

- The systems for detecting leakage from ponds, lagoons, and tanks are adequate to detect and ensure against any unplanned releases to ground water, surface water, or soil.

- The applicant controls and maintains releases to sewer systems to meet the requirements of 10 CFR 20.2003, "Disposal by Release into Sanitary Sewerage," including the following:

  - The material is water soluble.

  - Known or expected discharges meet the effluent limits specified in Table 3 of Appendix B to 10 CFR Part 20.

  - The known or expected total quantity of radioactive material released into the sewer system in a year does not exceed 5 curies (Ci) (185 gigabecquerels (GBq)) of hydrogen-3, 1 Ci (37 GBq) of carbon-14, and 1 Ci (37 GBq) of all other radioactive materials combined.

  Solubility is determined in accordance with the procedure described in NRC Information Notice 94-07.

- Reporting procedures comply with the requirements of 10 CFR 70.59 and the guidance in Regulatory Guide 4.16. The applicant provides reports that include the concentrations of principal radionuclides released to unrestricted areas in liquid and gaseous effluents and the MDC for the analysis and the error for each data point.

- The applicant's procedures and facilities for solid and liquid waste handling, storage, and monitoring result in safe storage and timely disposition of the material.

The scope of the applicant's environmental monitoring is acceptable if it is commensurate with the scope of activities at the facility and the expected impacts from operations as identified in the environmental report and if it meets the following criteria:

- Background and baseline concentrations of radionuclides in environmental media have been established through sampling and analysis.

- Monitoring includes sampling and analyses for monitoring air, surface water, ground water, soil, sediments, and vegetation, as appropriate.

- The description of monitoring identifies adequate and appropriate sampling locations and frequencies for each environmental medium, the frequency of sampling, and the analyses to be performed on each medium.

- Monitoring procedures employ acceptable analytical methods and instrumentation. The applicant commits to an instrument maintenance and calibration program that is appropriate to the given instrumentation. If the applicant proposes to use its own analytical laboratory for the analysis of environmental samples, the applicant commits to providing third-party verification of the laboratory's methods (such as that obtained by participation in a round-robin measurement program).

- Appropriate action levels and actions to be taken if the levels are exceeded are specified for each environmental medium and radionuclide.

  Action levels are selected on the basis of a pathway analysis that demonstrates that, below those concentrations, doses to the public will be ALARA *and* below the limits specified in Subpart B to 10 CFR Part 20. The action levels specify the concentrations at which an investigation would be performed and levels at which process operations would be shut down.

- MDCs are specified for sample analyses and are at least as low as those selected for effluent monitoring in air and water. MDCs for sediment, soil, and vegetation are selected on the basis of action levels to ensure that sampling and analytical methods are sensitive and reliable enough to support the application of the action levels.

- Data analysis methods and criteria that the applicant will use to evaluate and report the environmental sampling results are appropriate and will indicate when an action level is being approached in time to take corrective actions.

- The description of the status of all licenses, permits, and other approvals of facility operations required by Federal, State, and local authorities is complete and accurate.

- Environmental monitoring is adequate to assess impacts to the environment from potential radioactive and nonradioactive releases, as identified in high- and intermediate-consequence accident sequences in the ISA.

### 9.4.3.3 Integrated Safety Analysis Summary

In accordance with 10 CFR 70.60, "Applicability," applicants requesting a license to possess and process greater than a critical mass of SNM are required to perform an ISA and submit an ISA Summary to the NRC for approval. The applicant's treatment of environmental protection in the ISA is acceptable if it fulfills the following criteria:

- The ISA provides a complete list of accident sequences that result in radiological and nonradiological releases to the environment.

- The ISA uses acceptable methods to estimate environmental effects that may result from accident sequences and to determine whether the effects are high or intermediate consequences as defined in 10 CFR 70.61, "Performance Requirements." NUREG/CR-6410 describes acceptable methods for estimating environmental effects from accident sequences.

- The ISA provides a reasonable estimate of the likelihood and consequences of each accident sequence identified.

- The ISA identifies adequate engineering or administrative controls or both for each accident sequence of environmental significance. These controls will prevent or mitigate high- and intermediate-consequence accident sequences to an acceptable level. (Consequence categories are defined in 10 CFR 70.61 and in SRP Chapter 3.) IROFS provide the indicated level of protection.

- The ISA affords adequate levels of assurance so that IROFS will satisfactorily perform their safety functions. Configuration management, training, and maintenance activities contribute to achieving this assurance.

- For an ISA Summary of a facility that has not yet been constructed, the specifications for IROFS may not be complete at the time the ISA Summary is submitted. The IROFS functions should be described in sufficient detail for the reviewer to determine their adequacy to prevent or mitigate the accident sequence. For example, the description of an in-line gamma monitor used to alert an operator of an off-normal condition should define the range of gamma activity that the monitor needs to detect. A description of a ventilation system that controls the consequences of an enclosure spill should include its air-moving capacity. The description of a stack sampler that detects excessive airborne releases should include the capacities of the sampler.

In addition to participating in the integrated review of the ISA performed in accordance with SRP Chapter 3, the reviewer should also examine, in detail, the fire-initiated release scenarios provided in the ISA Summary to demonstrate compliance with 10 CFR 70.61 because fire-initiated accident scenarios have the potential for environmental consequences. This review should follow the guidance provided in applicable sections of SRP Chapter 3 to give a detailed evaluation of these scenarios, including a review of fire-induced consequences to the environment, the likelihood of such consequences, and IROFS chosen to prevent or mitigate those consequences.

The reviewer should consider the following factors in determining the acceptability of the applicant's descriptions of fire-initiated accident sequences:

- Scenario descriptions are sufficiently detailed to allow an understanding of the fire hazards that permits an evaluation of potential accident sequences.

- The applicant has adequately described the environmental consequences and likelihood of accident sequences identified in the ISA Summary involving fire, including risks from hazardous chemicals produced from licensed material and risks from radioactive materials.

- All controls that are used to mitigate or prevent the scenario are identified as IROFS or as defense-in-depth measures. For those controls that are IROFS, reliability and associated management measures should be indicated.

- Analyses that the applicant has performed as part of the evaluation should be part of the ISA and should be referenced or identified for potential further review by the NRC staff.

*9.4.3.4 Environmental Protection Management Measures*

The management measures applied to IROFS designated to prevent or mitigate accident sequences in which the IROFS are needed are acceptable if they meet the acceptance criteria in SRP Chapter 11.

## 9.5 Review Procedures

The staff will review the environmental report, environmental protection measures, ISA Summary, and management measures to verify that they meet the acceptance criteria defined in SRP Section 9.4. If the applicant has not provided sufficient information to make these determinations, the reviewer(s) will request additional information in coordination with the facility project manager. Chapter 4 of the NRC's Fuel Cycle Licensing Branch "Materials Licensing Procedures Manual," Revision 5, issued September 1996, specifies the format for a request for additional information.

### 9.5.1 Effluent and Environmental Controls and Monitoring

An environmental specialist will review the applicant's environmental protection measures in coordination with the fuel cycle facility inspector responsible for environmental protection. Any comments or concerns that the environmental reviewer or inspector identifies will be addressed and resolved, and the safety evaluation report (SER) (described in Section 9.6) for the licensing action will contain a statement indicating whether the inspection staff has any objections to the approval of the proposed licensing action. In addition, the review will include an evaluation of inspection reports and semiannual effluent reports, submitted in accordance with 10 CFR 70.59, to ensure licensee performance in environmental protection.

### 9.5.2 Integrated Safety Analysis Summary

As part of the environmental protection review, the environmental specialist will review the ISA Summary, including all identified accident sequences that can have significant environmental consequences, to determine whether the list completely and properly identifies all potential accidents. The environmental specialist will coordinate this review with the ISA reviewer. A detailed review will be conducted of (1) the accident sequences that, when left unmitigated, are rated as "high" consequence to an individual located outside the controlled area, (2) approximately 10 percent of the "intermediate" consequence sequences, and (3) a smaller number of accident sequences in which the consequences are less than intermediate. However, additional intermediate and low consequence sequences may be evaluated on the basis of the results of the initial review.

An evaluation of the ISA Summary requires coordination with other technical reviewers. The environmental review of the IROFS will be coordinated with the reviewers for the specific assurance functions, such as training and maintenance. The review of the ISA Summary may require the examination of the ISA and of detailed supporting ISA documents located at the facility. On the basis of these reviews, the reviewer should decide what supporting documents need to be reviewed. The reviewer will clearly identify in the SER either the materials examined and the descriptions and commitments considered and relied on or the basis for the staff's safety decision.

### 9.5.3 Management Measures

The environmental reviewer should review the grades assigned to the IROFS to ensure that the management measures described in SRP Chapter 11 are used to determine whether the licensee has established adequate management measures to apply to all IROFS designated to prevent or mitigate high or intermediate accident consequences to a member of the public or intermediate consequences to the environment in accordance with 10 CFR 70.61(b)(2), 10 CFR 70.61(c)(2), and 10 CFR 70.61(c)(3). The environmental reviewer should review the grades assigned to the IROFS to ensure that the management measures are graded commensurate with the reduction of risk attributable to the IROFS.

During the application and ISA Summary review, the reviewer should identify and communicate to the inspection staff any items or issues that should be inspected during an operational readiness review, if such a review will be performed. These items could include confirming that the engineered controls installed in the process actually meet the capabilities described in the ISA Summary and that administrative controls are implemented through procedures and operator training.

## 9.6 Evaluation Findings

An SER documents the evaluation findings of the environmental protection review of the application, including the review of the environmental protection program and the ISA Summary.

The staff reviewers will verify that the information submitted by the applicant is in accordance with 10 CFR Part 20, 10 CFR Part 51, and 10 CFR Part 70 and is consistent with the guidance in this SRP as it applies to environmental protection. In the input to the SER, the primary reviewer should document the bases for determining the adequacy of the application with respect to environmental protection and should recommend additional license conditions in areas where the application is not adequate. The primary reviewer also describes the applicant's approach to ensuring the quality and reliability of the IROFS required for environmental protection.

Environmental protection is often reviewed and evaluated in conjunction with the environmental report, and the EA or EIS summarizes the environmental protection function. However, the EA or EIS does not become part of the license. The SER should briefly discuss issues identified during the review, and any recommended license conditions based on the analysis in the EA or EIS should be added to the license.

If an EA or EIS is prepared for the licensing action, the environmental safety section of the SER should report the date the document was issued. If the EA results in a FONSI, the SER should include the publication date of the FONSI in the *Federal Register*. If an EIS is prepared, the SER would include the *Federal Register* publication date for the record of decision. When applicable, the SER will also document the determination that an action meets the requirements for a categorical exclusion.

### 9.6.1 Safety Evaluation

The following language would be appropriate for a licensing action that requires an environmental review:

> The applicant has committed to adequate environmental protection measures, including (1) environmental and effluent monitoring and (2) effluent controls to maintain public doses ALARA as part of the radiation protection program. The NRC staff concludes, with reasonable assurance, that the applicant's conformance to the application and license conditions is adequate to protect the environment and public health and safety and to comply with the regulatory requirements imposed by the Commission in 10 CFR Part 20, 10 CFR Part 51, and 10 CFR Part 70. The bases for these conclusions are as follows:
>
> [State the bases for the conclusion, including any recommended license conditions.]

If the action requires preparation of an EIS, the SER should include the following language:

> The NRC staff prepared an environmental impact statement (EIS) [publication date] for this licensing action as required by 10 CFR 51.20. On the basis of the EIS, the NRC stated in its record of decision [publication date in the *Federal Register*] that the preferred option was [state preferred option here].

If the action requires preparation of an EA and results in a FONSI, the SER should include the following language:

> The NRC staff prepared an environmental assessment (EA) for this action as required by 10 CFR 51.45 and 10 CFR 51.60. On the basis of the EA, the staff has reached a finding of no significant impact, published in the *Federal Register* on [publication date and FR citation].

If the staff determines that the action was categorically excluded from environmental review under 10 CFR 51.22, the SER should include the following language:

> The staff has determined that the amended actions are administrative, organizational, or procedural in nature. Based on this evaluation, there is no significant impact to the environment, and the action of amending the license is eligible for categorical exclusion. Therefore, in accordance with 10 CFR 51.22(c)(11), neither an environmental assessment nor an environmental impact statement is required for this action. The regulation at 10 CFR 51.22(c)(11) allows for a categorical exclusion if the following four requirements have been satisfied:
>
> (1)    There is no significant change in the types or significant increase in the amounts of any effluents that may be released off site.
>
> (2)    There is no significant increase in individual or cumulative occupational radiation exposure.
>
> (3)    There is no significant construction impact.

(4)     There is no significant increase in the potential for, or consequences from, radiological accidents.

The changes made in this licensing action do not pose a significant change or increase in parameters (1) through (4) above. There are no changes in the types or increases in the amounts of effluents. Occupational exposure is expected to remain the same. These changes involve no additional construction activity. The potential for, and consequences from, radiological accidents are expected to be the same.

[State the bases for the conclusion, including any recommended license conditions.]

The NRC staff prepared an environmental impact statement (EIS) [publication date] for this licensing action as required by 10 CFR 51.20. On the basis of the EIS, the NRC stated in its record of decision [publication date in the *Federal Register*] that the preferred option was [state preferred option here].

## 9.7  References

*U.S. Code of Federal Regulations*, Chapter I, Title 10, "Energy," Part 20, "Standards for Protection against Radiation."

*U.S. Code of Federal Regulations*, Chapter I, Title 10, "Energy," Part 51, "Environmental Protection Regulations for Domestic Licensing and Related Regulatory Functions."

*U.S. Code of Federal Regulations*, Chapter I, Title 10, "Energy," Part 70, "Domestic Licensing of Special Nuclear Material."

American National Standards Institute, "Guide to Sampling Airborne Radioactive Materials in Nuclear Facilities," ANSI N13.1-1982.

American National Standards Institute, "Specification and Performance of On-Site Instrumentation for Continuously Monitoring Radioactive Effluents," ANSI N42.18-1980.

National Council on Radiation Protection and Measurements, "Screening Models for Releases of Radionuclides to Atmosphere, Surface Water, and Ground," NCRP Report No. 123, Volumes I and II, January 1996.

U.S. Environmental Protection Agency, "Limiting Values of Radionuclide Intake and Air Concentration and Dose Conversion Factors for Inhalation, Submersion, and Ingestion," Federal Guidance Report No. 11, September 1988.

U.S. Nuclear Regulatory Commission, "Guidance to Hazardous, Radioactive, and Mixed Waste Generators on the Elements of a Waste Minimization Program," Information Notice No. 94-23, March 25, 1994.

U.S. Nuclear Regulatory Commission, "Solubility Criteria for Liquid Effluent Releases to Sanitary Sewerage under the Revised 10 CFR Part 20," Information Notice 94-07, January 28, 1994.

U.S. Nuclear Regulatory Commission, NMSS/FCSS/Fuel Cycle Licensing Branch, "Materials Licensing Procedures Manual," Revision 5, September 1996.

U.S. Nuclear Regulatory Commission, "Quality Assurance for Radiological Monitoring Programs (Inception through Normal Operations to License Termination)—Effluent Streams and the Environment," Regulatory Guide 4.15, Revision 2, July 2007.

U.S. Nuclear Regulatory Commission, "Monitoring and Reporting Radioactivity in Releases of Radioactive Materials in Liquid and Gaseous Effluents from Nuclear Fuel Processing and Fabrication Plants and Uranium Hexafluoride Production Plants," Regulatory Guide 4.16, Revision 1, December 1985.

U.S. Nuclear Regulatory Commission, "Constraint on Releases of Airborne Radioactive Materials to the Environment for Licensees Other Than Power Reactors," Regulatory Guide 4.20, December 1996.

U.S. Nuclear Regulatory Commission, "ALARA Levels for Effluents from Materials Facilities," Regulatory Guide 8.37, July 1993.

U.S. Nuclear Regulatory Commission, "Nuclear Fuel Cycle Accident Analysis Handbook," NUREG/CR-6410, March 1998.

U.S. Nuclear Regulatory Commission, "Integrated Safety Analysis Guidance Document," NUREG-1513, May 2001.

U.S. Nuclear Regulatory Commission, "Environmental Review Guidance for Licensing Actions Associated with NMSS Programs," NUREG-1748, August 2003.

# 10. DECOMMISSIONING

## 10.1 Purpose of Review

The purpose of the review of the applicant's decommissioning plans is to determine with reasonable assurance that the applicant will be able to decommission the facility safely and in accordance with the requirements of the U.S. Nuclear Regulatory Commission (NRC).

At the time of the initial application and at license renewal, the applicant or licensee should discuss its conceptual approach for meeting the decommissioning requirements in Title 10 of the *Code of Federal Regulations* (10 CFR) Part 20, "Standards for Protection against Radiation," Subpart E, "Radiological Criteria for License Termination." The applicant or licensee should discuss its plans for minimizing contamination.

At the time of the initial license application and again at license renewal, the applicant or licensee may be required to submit a decommissioning funding plan (DFP) in accordance with 10 CFR 70.25(b). The purpose of the NRC's evaluation of the DFP is to determine whether the applicant or licensee has taken the following actions:

* considered decommissioning activities that may be needed in the future

* performed a credible site-specific cost estimate for those activities

* presented the NRC with financial assurance to cover the cost of those activities in the future

Therefore, the DFP should contain the following:

* an overview of the proposed decommissioning activities
* the methods used to determine the cost estimate
* the financial assurance mechanism

This overview must contain sufficient detail to enable the reviewer to determine whether the decommissioning cost estimate is reasonably accurate.

In the application, the applicant or licensee should discuss its plans for meeting the decommissioning recordkeeping requirements in 10 CFR 70.25(g). Under the regulations, a licensee must keep records important for decommissioning. These records include records of spills or unusual occurrences involving the spread of contamination, as-built drawings and modifications to structures and equipment in restricted areas, a list of areas designated or formerly designated as restricted areas, and records pertaining to the financial assurance requirements.

If required by 10 CFR 70.38(g), the licensee must also submit, for NRC approval, a decommissioning plan (DP) before beginning its decommissioning actions. The DP must detail the specific decommissioning activities that the licensee will perform and describe the radiation protection procedures that the licensee will use to protect workers, the public, and the environment during decommissioning.

This information must be sufficient to enable the reviewer to assess the appropriateness of the decommissioning activities and the adequacy of the procedures to protect the health and safety of workers, the public, and the environment. It must also update the cost estimate originally presented in the DFP to undertake the facility decommissioning. The licensee can generally obtain approval of a DP by submitting an application for a license amendment. The reviewer must determine that the applicant understands the decommissioning requirements and procedures and that it commits to protecting the health and safety of workers, the public, and the environment during decommissioning.

## 10.2 Responsibility for Review

Primary:       Health Physics Reviewer

Secondary:     Environmental Reviewer
               Technical and Financial Specialists in the Division of Waste Management and
                   Environmental Protection

Supporting:    Fuel Facility Inspection Staff

## 10.3 Areas of Review

Before beginning to review the DFP or DP, the reviewer should first evaluate the applicant's proposed environmental protection measures (see Chapter 9 of this Standard Review Plan (SRP)) and, specifically, the commitments to minimize waste associated with decommissioning. In addition, the reviewer should evaluate the applicant's radiation protection program (see SRP Chapter 4) as it applies to radiological decontamination and the management of radiological effluents.

The staff review should cover the following areas:

- conceptual decontamination and decontamination plan, including the decommissioning program and steps, management and organization, health and safety, radiological decommissioning criteria, waste management, security and nuclear material control, recordkeeping, the decontamination process, and the minimization of contamination

- decommissioning costs and financial assurance (i.e., the decommissioning cost information should be consistent with the recommendations in NUREG-1757, "Consolidated Decommissioning Guidance," issued September 2006)

The reviewer will evaluate the applicant's DFP or DP or both in accordance with NUREG-1757.

## 10.4 Acceptance Criteria

### 10.4.1 Regulatory Requirements

The following NRC regulations require planning, financial assurance, and recordkeeping for decommissioning, as well as procedures and activities to minimize waste and contamination:

- 10 CFR Part 20, Subpart E, "Radiological Criteria for License Termination"

- 10 CFR 30.35, "Financial Assurance and Recordkeeping for Decommissioning"

- 10 CFR 30.36, "Expiration and Termination of Licenses and Decommissioning of Sites and Separate Buildings or Outdoor Areas"

- 10 CFR 40.14, "Specific Exemptions"

- 10 CFR 40.36, "Financial Assurance and Recordkeeping for Decommissioning"

- 10 CFR 40.42, "Expiration and Termination of Licenses and Decommissioning of Sites and Separate Buildings or Outdoor Areas"

- 10 CFR 70.17, "Specific Exemptions"

- 10 CFR 70.22(a)(9)

- 10 CFR 70.25, "Financial Assurance and Recordkeeping for Decommissioning"

- 10 CFR 70.38, "Expiration and Termination of Licenses and Decommissioning of Sites and Separate Buildings or Outdoor Areas"

### 10.4.2 Regulatory Guidance

NUREG-1757 is relevant to the decommissioning of fuel cycle facilities.

## 10.5 Review Procedures

The primary reviewer will evaluate the application against the NRC requirements and acceptance criteria identified in NUREG-1757. A detailed review of any contamination and waste minimization plans that the applicant submits in response to 10 CFR 20.1406, "Minimization of Contamination," will supplement this review (as appropriate). The reviewer will also coordinate with the principal reviewers for environmental protection listed in SRP Chapter 9 to confirm the review of a new applicant's plans to minimize waste and the plans for existing licensees to minimize contamination and reduce exposures and effluents as part of the radiation protection program established under 10 CFR Part 20. The purpose of this coordination is to ensure that any issues that are relevant to the environmental review are properly conveyed to the primary reviewer for consideration and resolution as part of the review discussed in SRP Chapter 9 and that any decommissioning issues that arise in the environmental review that are best suited for review using guidance in this chapter are conveyed to the primary reviewer for consideration and resolution.

If the decommissioning review identifies the need for the applicant to submit information that has not already been included in the application, the reviewer will document these additional information needs in a request for additional information (RAI). The RAI transmitted to the applicant will specify a reasonable amount of time (e.g., 30 to 60 days) for the applicant to reply. Failure of the applicant to provide the requested information by the specified date, or on an alternative schedule that is mutually agreeable, could be grounds for terminating or suspending the application review.

The primary reviewer should coordinate with the Division of Waste Management and Environmental Protection to obtain the appropriate technical assistance in reviewing proposed DPs and financial assurance measures. The primary reviewer will coordinate with reviewers assigned by the Division of Waste Management and Environmental Protection to incorporate, as appropriate, RAIs and review findings in licensing correspondence and safety evaluation reports (SERs) related to decommissioning.

The reviewer should perform a safety review using the acceptance criteria in NUREG-1757 to ensure that the proposed decommissioning methodology, principal remediation activities, and worker and environmental radiation protection programs are acceptable.

## 10.6 Evaluation Findings

If the applicant provides sufficient information to satisfy the acceptance criteria and requirements identified in Section 10.4, the staff will conclude that the DFP or DP evaluation is complete and satisfactory. The primary reviewer will prepare an SER for the licensing project manager in support of the licensing action. This SER should address each topic area reviewed and discuss why the NRC has reasonable assurance that the DFP or DP should be considered acceptable, explaining the bases for the reviewers' conclusions. The SER may also include license conditions where the application is deficient. The SER should include a summary statement of what the staff evaluated. As an example, the staff might document its evaluation in an SER, using the following language:

> The NRC staff has evaluated the applicant's/licensee's plans and financial assurance for decommissioning in accordance with NUREG-1757, "Consolidated Decommissioning Guidance," issued September 2006. On the basis of this evaluation, the NRC staff has determined that the applicant's/licensee's plans and financial assurance for decommissioning comply with the NRC's regulations and provide reasonable assurance of protection for workers, the public, and the environment.

## 10.7 References

*U.S. Code of Federal Regulations*, Chapter I, Title 10, "Energy," Part 20, "Standards for Protection against Radiation."

*U.S. Code of Federal Regulations*, Chapter I, Title 10, "Energy," Part 30, "Rules of General Applicability to Domestic Licensing of Byproduct Material."

*U.S. Code of Federal Regulations*, Chapter I, Title 10, "Energy," Part 40, "Domestic Licensing of Source Material."

*U.S. Code of Federal Regulations*, Chapter I, Title 10, "Energy," Part 70, "Domestic Licensing of Special Nuclear Material."

U.S. Nuclear Regulatory Commission, "Consolidated Decommissioning Guidance," NUREG-1757, September 2006.

# 11. MANAGEMENT MEASURES

## 11.1 Purpose of Review

Management measures are activities performed by a licensee, generally on a continuing basis, that are applied to IROFS to provide reasonable assurance that the items relied on for safety (IROFS) will perform their intended safety function when needed to prevent accidents or mitigate the consequences of accidents to an acceptable level. The purpose of management measures is to provide reasonable assurance of compliance with Title 10 of the *Code of Federal Regulations* (10 CFR) 70.61, "Performance Measures." Reasonable assurance is established by considering factors such as necessary maintenance, operating limits, common-cause failures, and the likelihood and consequences of failure or degradation of the IROFS and the measures. As defined in 10 CFR 70.4, "Definitions," management measures include configuration management (CM), maintenance, training and qualification, procedures, audits and assessments, incident investigations, records management, and other quality assurance (QA) elements.

## 11.2 Responsibility for Review

Primary:     Quality Assurance Reviewer

Supporting:  Primary Reviewers of Chapters 3 through 10 of this Standard Review Plan (SRP)
             Fuel Cycle Facility Inspector

## 11.3 Areas of Review

According to 10 CFR 70.62(d), each applicant must establish management measures to ensure that IROFS, as documented in the integrated safety analysis (ISA) summary, provide reasonable assurance that they will be designed, implemented, and maintained in such a way as to ensure that they are available and reliable to perform their intended functions, when needed, to comply with the performance requirements of 10 CFR 70.61. The degree to which measures are applied may be a function of the item's importance in meeting the performance requirements. If a "graded" application of a particular management measure is used for IROFS of differing importance, the applicant should describe the variations and the reviewer should determine whether the measures are commensurate with the importance to safety of the IROFS.

The specific areas of review are as follows:

*   Configuration Management—The U.S. Nuclear Regulatory Commission (NRC) staff review will determine whether the applicant has proposed a CM program that ensures consistency in the facility design and operational requirements, the physical configuration, and the facility documentation. The review should determine that the applicant's CM program captures formal documentation governing the design and continued modification of the site, structures, processes, systems, components, computer programs, personnel activities, and supporting management measures. The review should also ensure that the CM program is adequately coordinated and integrated with other management measures.

The NRC staff should review the applicant's descriptions and commitments for CM, including descriptions of the organizational structure responsible for CM activities; descriptions of the process, procedures, and documentation required by the applicant for modifying the site; and descriptions of the various levels of CM to be applied to IROFS designated in the ISA Summary. The staff's review should focus on the applicant's CM measures that provide reasonable assurance of the documentation of engineering, procurement, installation, and modifications; the training and qualification of affected staff; the revision and distribution of operating, test, calibration, surveillance, and maintenance procedures and drawings; and the postmodification testing. The review of the overall approach to implementing CM should include the evaluation of the CM program, design requirements, document control, change control, assessments, and design reconstitution for existing facilities.

- Maintenance—The NRC staff's review should evaluate the applicant's description of its maintenance program. The staff will examine the applicant's commitments to inspect, calibrate, test, and maintain IROFS to a level commensurate with the items' importance to safety. The staff will review the applicant's description of how the site organization implements (1) corrective maintenance, (2) preventive maintenance (PM), (3) surveillance and monitoring, and (4) functional testing. Not every aspect of each of the four maintenance functions is necessarily required. The applicant should justify the assignment of differing degrees of maintenance to individual IROFS based on the item's contribution to risk reduction.

- Training and Qualification—The regulations at 10 CFR Part 70, "Domestic Licensing of Special Nuclear Material," require that all personnel who perform activities relied on for safety be trained and tested so as to provide reasonable assurance that they understand, recognize the importance of, and are qualified to perform these activities in a manner that adequately protects public health and safety and the environment. As appropriate for their authority and responsibilities, these personnel should have the knowledge and skills necessary to design, operate, and maintain the facility safely. Therefore, the application should describe the training, testing, and qualification of these personnel, and the NRC staff should review this description. The review should examine the applicant's experience and capabilities to provide this training for its personnel who will perform activities relied on for safety. The review of the training and qualification should address the following areas:

  – organization and management of the training function

  – analysis and identification of functional areas requiring training

  – position training requirements

  – development of the basis for training, including objectives

  – organization of instruction and use of lesson plans and other training guides

  – evaluation of trainee learning

  – conduct of on-the-job training

  – evaluation of training effectiveness

- personnel qualification

- provisions for continuing quality assurance, including the needs for retraining or reevaluation of qualification

- Procedures—The NRC staff review should examine the applicant's process for the preparation, use, and management control of written procedures. This process should include the basic elements of identification, development, verification, review and comment resolution, approval, validation, issuance, change control, and periodic review.

- The actual operating procedures are not part of the license, and the NRC staff would not normally review them for technical adequacy since the inspection function addresses this aspect. The NRC staff should review the license application to ensure that the applicant's process for establishing procedures adequately addresses the following areas:

  - method for identifying procedures that are needed plantwide

  - essential elements that are generic to all procedures

  - method for creating and controlling procedures within plant management control systems

  - method for verifying and validating procedures before use

  - method for periodically reverifying and revalidating procedures

  - method for ensuring that current procedures are available to personnel and those personnel are qualified to use the latest procedure

- Audits and Assessments—The NRC staff should review the applicant's program of audits and assessments. The program should consist of two distinct levels of activities: (1) an audit activity structured to monitor compliance with regulatory requirements and license commitments and (2) an assessment activity oriented to determining the effectiveness of the activities in achieving applicant-specified objectives that provide reasonable assurance of the continued availability and reliability of IROFS. An applicant may describe a corrective action program, which includes the functions of incident investigations as well as audits and assessment. This approach is acceptable and the reviewer should, in that case, review the applicant's description and commitments with regard to the acceptance criteria in this SRP chapter for incident investigations as well as audits and assessments. The review of the audits and assessments should address the following areas:

- the commitments to audit and assessment activities

- the use of qualified and independent audit and assessment personnel

- the general structure of typical audits and assessments

- the facility procedures to be used to direct and control the audit and assessment activities

- the planned use of the results of the audit and assessment activities

- the documentation to record and distribute the findings and recommendations of these audits and assessments

- the planning and implementation of corrective actions based on the findings and recommendations

- Incident Investigations—The NRC staff should review the applicant's program, procedures, and management structure for investigating abnormal events and completing appropriate corrective actions. The review should include the provisions for establishing investigating teams, the methods for determining root causes, and the procedures for tracking and completing corrective actions and for documenting the process for the purpose of applying the "lessons learned" to other operations. The applicant may describe a corrective action program, which includes the functions of audits and assessment as well as incident investigations. This approach is acceptable and the reviewer should, in that case, review the applicant's description and commitments with regard to the acceptance criteria in this SRP chapter for audits and assessments as well as incident investigations.

- Records Management—The requirements for the management of records vary according to the nature of the facility and the hazards and risks it poses. The staff should review areas related to the handling and storage of health and safety records and the records generated or needed in the design, construction, operation, and decommissioning phases of the facility. The review should provide reasonable assurance that the records management function is adequately coordinated and integrated with other management measures. The staff should review the following:

  - the process whereby records (i.e., training records, dosimetry records, effluents records, records of classified information, records concerning facility IROFS and their failures) are created, selected, verified, categorized, indexed, inventoried, protected, stored, maintained, distributed, deleted, or preserved

  - the handling and control of various kinds of records (including contaminated and classified records) and the media in which the records are captured

  - the physical characteristics of the records storage area(s) with respect to the preservation and protection of the records for their designated lifetimes

- Other Quality Assurance Elements—The application should address other QA elements that will be applied to IROFS and other management measures. The NRC staff should evaluate whether the application of other QA elements is adequately described. The staff's review objective is to obtain reasonable assurance that the design, procurement, construction, operation, maintenance, inspection, testing, and modification phases of a facility's life cycle implement accepted QA principles. The NRC staff should examine the applicant's commitment to overall QA, the selection of quality criteria and quality level, and the proposed method for implementation. Application of graded QA and quality levels commensurate with the risk involved should parallel the same risk levels established for maintenance and other management measures.

The reviewer should recognize that facility safety may not be the only area at a fuel cycle facility requiring QA elements. The applicant's customers and the NRC, under 10 CFR Part 50, "Domestic Licensing of Production and Utilization Facilities," may impose product-related QA criteria. This SRP limits the focus of the review of QA measures to ensuring the safety of workers and the public and protecting the environment (i.e., in relation to the performance requirements of 10 CFR 70.61).

## Review Interfaces

Other sections of the license application may include information on CM, maintenance, training and qualification, procedures, audits and assessments, incident investigations, record management, or other QA elements applied to management measures. The NRC staff should focus its review activities on management measures associated with IROFS of high risk importance. The reviewer of this SRP chapter should coordinate with the reviewers of SRP Chapters 3 through 10 to inform the selection of management measures for more detailed review.

## 11.4 Acceptance Criteria

### 11.4.1 Regulatory Requirements

Acceptance criteria are based on meeting the relevant requirements of the regulations described in this section.

The regulatory basis for the review is 10 CFR 70.22, "Contents of Applications," and 10 CFR 70.65, "Additional Content of Applications." In addition, the management measures review should provide reasonable assurance of compliance with the following regulations:

- 10 CFR 70.4 states that management measures include CM, maintenance, training and qualification, procedures, audits and assessments, incident investigations, records management, and other QA elements.

- 10 CFR 70.22(a)(8) requires that each application for a license must contain proposed procedures to protect health and minimize danger to life or property.

- 10 CFR 70.62(a)(3) states that records must be kept for all IROFS failures, describes required data to be reported, and sets time requirements for updating the records.

- 10 CFR 70.62(d) requires an applicant to establish management measures for engineered and administrative controls and control systems that are identified as IROFS, in accordance with 10 CFR 70.61(e), so that they are available and reliable to perform their functions when needed.

- 10 CFR 70.64(a)(1) states that new facilities or new processes at existing facilities must develop and implement designs, in accordance with management measures, to provide reasonable assurance that IROFS will be designed, implemented, and maintained to ensure that they are available and reliable to perform their safety function when needed.

- 10 CFR 70.64(a)(1) states that the licensee must maintain or control appropriate records of IROFS throughout the life of the facility.

- 10 CFR 70.64(a)(8) states that the design of IROFS must provide for inspection, testing, and maintenance adequate to ensure their availability and reliability to perform their function when needed.

- Facility change and change processes must conform to 10 CFR 70.72, "Facility Changes and Change Process."

- 10 CFR 70.74(a) and 10 CFR 70.74(b) require incident investigation and reporting.

### 11.4.2 Regulatory Guidance

Regulatory guidance appears in American National Standards Institute/American Society of Mechanical Engineers (ANSI/ASME) NQA-1-1983, "Quality Assurance Requirements for Nuclear Facility Applications," as endorsed by Regulatory Guide 1.28, "Quality Assurance Program Requirements (Design and Construction)," Revision 3, issued August 1985. This guidance applies only to applications for plutonium processing and fuel fabrication facilities.

### 11.4.3 Regulatory Acceptance Criteria

The reviewer should find the applicant's management measures acceptable if the applicant has met the acceptance criteria described in the following sections or has identified and justified an alternative approach.

#### 11.4.3.1 Configuration Management

The regulation at 10 CFR 70.4 defines CM as a management measure that provides oversight and control of design information, safety information, and records of modifications that might impact the ability of IROFS to perform their functions when needed. The applicant's description of CM is acceptable if it meets the following conditions:

- The application describes the CM program, design requirements, document control, change control, assessments, and design reconstitution (for existing facilities only).

- The application describes the CM program and defines the specific attributes of the levels of CM that will be applied to select IROFS.

- The ISA Summary clearly defines the IROFS to be listed under CM along with the assignment of any grades or quality levels. The applicant should indicate in the ISA Summary the level of CM attributes that is applied to a particular IROFS. However, in the ISA Summary, this indication may consist of only an index or category designation.

- The application describes a design process leading to drawings and other statements of requirements that proceeds logically from the design basis.

- The application describes how design requirements and associated design bases are established and are maintained through control of the design process. It also describes technical management review and approval functions.

- The application describes an acceptable method to create and control documents that are relied on for safety. These documents include design requirements, ISAs, as-built drawings, specifications, all procedures that are IROFS, procedures involving training, QA, maintenance, audits and assessments, emergency operating procedures, emergency response plans, system modification documents, assessment reports, and others that the applicant deems part of CM.

- The application describes how the CM function will maintain strict consistency among the design requirements, the physical configuration, and the facility documentation.

- The application contains a commitment to evaluate, implement, and track each change to the site, structures, processes, systems, equipment, components, computer programs, and activities of personnel.

- The application describes an acceptable process for providing reasonable assurance that the ISA is systematically reviewed and modified to reflect design or operational changes from an established safety basis and that all documents outside the ISA that are affected by safety-basis changes are properly modified, authoritatively approved, and made available to personnel.

- The application describes the documentation process following changes made in accordance with 10 CFR 70.72. Changes to the affected onsite documentation should be made promptly to avoid inadvertent access by facility personnel to outdated design and other specifications for IROFS.

- The application confirms that initial and periodic assessments of the CM function are conducted to determine the program's effectiveness and to correct deficiencies. The application indicates that such assessments are systematically planned and conducted in accordance with an overall facility audit and assessment function.

- For existing facilities, the application may describe whatever design reconstitution has been done for the purpose of the application. The applicant has available the current design bases, including design requirements, supporting analyses, and documentation supporting all IROFS.

- For new facilities or new processes at existing facilities, the application describes facility and system design and facility layout based on defense-in-depth practices, in accordance with 10 CFR 70.64, "Requirements for New Facilities or New Processes at Existing Facilities." Defense-in-depth practices should be applied early through the completion of design by providing successive levels of protection such that health and safety will not wholly depend on any single element of the design, construction, maintenance, or operation of the facility.

## 11.4.3.2 Maintenance

As required by 10 CFR 70.62(d), engineered and administrative controls that are identified as IROFS must be designed, implemented, and maintained to ensure that they are available and reliable when needed.

The regulation at 10 CFR 70.64(a)(8) requires that IROFS for new facilities or new processes at existing facilities receive adequate inspection, testing, and maintenance to ensure their availability and reliability when needed.

The reviewers should find the applicant's submittal acceptable if the application includes the following information:

- descriptions of corrective maintenance, PM, surveillance and monitoring, and functional testing

- description of how the maintenance function will be designed to ensure that the objective of preventing failures through maintenance is appropriately balanced against the objective of minimizing unavailability of IROFS because of monitoring or PM

- discussion of how the maintenance function uses, interfaces with, or is linked to the various management measures

- justifications for assignment of differing degrees of maintenance to individual IROFS, based on the item's contribution to the reduction of risk

- for IROFS identified in the ISA Summary, description of the surveillance function and its conduct at a specified frequency

- description of how the surveillance activity supports the determination of performance trends for IROFS, thus providing data useful in determining PM frequencies

- description of the applicant's retention of records of the current surveillance schedule, performance criteria, and test results for all IROFS

- for surveillance tests that can be done only while IROFS are out of service, description of the proper compensatory measures that will be prescribed for the continued normal operation of a process

- description of how the results of incident investigations, the review of failure records required by 10 CFR 70.62(a)(3), and identified root causes are used to modify the affected maintenance function and eliminate or minimize the root cause

- documentation of the approach to performing corrective actions or repairs on IROFS

- description of how the maintenance function provides a planned, systematic, integrated, and controlled approach for the repair and replacement activities associated with identified unacceptable performance deficiencies of IROFS

- description of the PM function that demonstrates a commitment to conducting preplanned and scheduled periodic refurbishing, or partial or complete overhaul, of IROFS to minimize occurrences of their unanticipated losses

- description of the applicant's retention of records showing the PM schedule and results for all IROFS subject to this maintenance component

- general description of the methods used and the commitment to perform functional testing, as needed, of IROFS after PM or corrective maintenance

- as necessary, a commitment to conduct functional tests designed to include all operational aspects of the IROFS that are important to safety during startup of new processes

- description of how the applicant will maintain records showing the functional test schedule and results for all IROFS subject to this maintenance component

- general discussion of how the applicant will verify that the administrative controls identified as IROFS are available and reliable to perform their intended safety function over extended periods of operation

### 11.4.3.3 Training and Qualification

The application should be acceptable regarding personnel training and qualification if it satisfies the criteria described below. In addition to the regulatory acceptance criteria, the SRP provides additional specific criteria for (1) training and qualification for radiation safety personnel in Section 4.4.5, (2) criticality safety in Section 5.4.3.2, and (3) emergency planning in Section 8.4.3.1.11. Similarly, some of the information specified below may appear in other sections of the application and may be incorporated by reference. Review criteria include the following:

- The application should include the following commitments regarding organization and management of training:

  - Line management is responsible for the content and effective conduct of the training.

  - The application clearly defines job function, responsibility, authority, and accountability of personnel involved in managing, supervising, and implementing training.

  - The applicant uses performance-based training as the primary management tool for analyzing, designing, developing, conducting, and evaluating training.

  - The applicant documents and implements procedures to provide reasonable assurance that all phases of training are conducted reliably and consistently.

  - The applicant ensures that training documents are linked to the CM system to provide reasonable assurance that the training reflects design changes and modifications.

- The applicant grants exemptions from training to trainees and incumbents only when justified, documented, and approved by management.

- The applicant maintains both programmatic and individual training records. These records support management information needs and provide required data on each individual's training and qualification.

- The applicant should provide formal training for each position or activity that is relied on for safety. Training may be classroom or on-the-job training or both. The application should state what training will be conducted and which personnel will be required to complete it. The application should also demonstrate the following:

  - The applicant ensures that each activity selected for training (initial or continuing) from the facility-specific activities is correlated with supporting procedures and training materials.

  - The applicant reviews facility-specific activities selected for training and compares training materials on an established schedule, updating them as necessitated by changes in procedures, facility systems and equipment, or job scope. The applicant monitors and evaluates change actions (e.g., procedure changes, equipment changes, facility modifications) for their impact on the development or modification of initial and continuing training and incorporates such change actions in a timely manner.

- The application should contain commitments regarding personnel qualification for managers, supervisors, designers, technical staff, construction personnel, facility operators, technicians, maintenance personnel, and other staff who perform regulated activities.

- The application should contain commitments regarding minimum qualifications for personnel. Minimum qualifications should be commensurate with the assigned functional responsibility and authority of the respective personnel as detailed below:

  - Managers should have a bachelor of science (B.S.), bachelor of art (B.A.), or equivalent degree. Each manager should have either management or technical experience in a facility similar to the facility identified in the application.

  - Supervisors should have at least the qualifications required of personnel being supervised.

  - Technical professional staff whose actions or judgments are critical to satisfying the performance requirements identified in 10 CFR Part 70 should have a B.S., B.A., or equivalent degree in the appropriate technical field.

  - Construction personnel, facility operators, technicians, maintenance personnel, and other staff whose actions are required to comply with NRC regulations should have completed the applicant's training process or have equivalent experience or training.

–   The applicant should require candidates for process operators to meet the minimum qualifications described in the application. The applicant should require candidates for job functions other than process operators to meet minimum qualifications, but the application need not describe these minimum qualifications.

- Training objectives should state the knowledge, skills, and abilities that the trainee should acquire; the conditions under which required actions will take place; and the standards of performance the trainee should achieve upon completion of the training activity.

- Lesson plans and other training guides should provide guidance to ensure the consistent conduct of training activities and should be based on required learning objectives derived from specific job performance requirements.

- The applicant should use lesson plans or guides for all training, and these lesson plans or guides should include standards for evaluating acceptable trainee performance. The evaluation of trainee accomplishment is acceptable if the applicant evaluates trainees periodically during training to determine their progress toward full capability to perform the job requirements and at the completion of training to determine their capability to perform the job requirements.

- The applicant should establish review and approval requirements for all lesson plans or guides and other training materials before their issue and use.

- The application should describe any on-the-job training used for activities relied on for safety.

- The applicant should conduct on-the-job training using well-organized and current training materials. Designated personnel who are competent in the program standards and training methods should conduct the training.

- Completion of on-the-job training should be by actual task performance. When the actual task cannot be performed and is, therefore, "walked down," the conditions of task performance, references, tools, and equipment should reflect the actual task to the extent possible.

- Provisions for continuing assurance of personnel training and qualification are acceptable if the application addresses periodic requalification of personnel by training or testing or both, as necessary, to provide reasonable assurance that personnel continue to understand, recognize the importance of, and be qualified to perform activities that are relied on for safety.

- An evaluation of training effectiveness and its relation to job performance is acceptable if it provides reasonable assurance that the training conveys all required skills and knowledge and is used to revise the training, where necessary, based on the performance of trained personnel in the job setting. The application should also demonstrate the following:

- Qualified individuals should periodically conduct a comprehensive evaluation of individual training to identify strengths and weaknesses. The applicant should use feedback from trainee performance during training and from former trainees and their supervisors to evaluate and refine the training.

- The applicant should initiate, evaluate, track, and incorporate improvements and changes to initial and continuing training to correct training deficiencies and performance problems.

### 11.4.3.4 Procedures

The regulation at 10 CFR 70.22(a)(8) requires that the application contain procedures to protect public health and safety. The application is acceptable in this regard if it describes the applicant's process for developing and implementing procedures and satisfies the following:

- The applicant provides information regarding the procedure categories used at the facility. The categories typically include management control, operating, maintenance, and emergency procedures.

- The applicant writes or plans procedures for the operation of IROFS and for all management measures supporting those IROFS.

- The applicant includes the following commitment regarding procedure adherence: "Activities involving licensed SNM and/or IROFS will be conducted in accordance with approved procedures."

- The applicant develops procedures for sitewide safe work practices to control processes and operations with licensed special nuclear material (SNM) and/or IROFS and/or hazardous chemicals incident to the processing of licensed material.

- The applicant has existing or planned procedures to direct the following activities: (1) design, (2) CM, (3) procurement, (4) construction, (5) radiation safety, (6) maintenance, (7) QA elements, (8) training and qualification, (9) audits and assessments, (10) incident investigations, (11) records management, (12) criticality safety, (13) fire safety, (14) chemical process safety, and (15) reporting requirements.

- Procedures are required for operator actions that are necessary to prevent or mitigate accidents identified in the ISA Summary. The applicant provides a listing of the types of activities that are covered, or are planned to be covered, by written procedures. The listing includes the topics of administrative procedures; system procedures that address startup, operation, and shutdown; abnormal operation or alarm response; maintenance activities that address system repair, calibration, inspection, and testing; and emergency procedures. Appendix A to this SRP chapter provides an acceptable listing of the items to be included under each topic.

- The applicant describes the method for identifying, developing, approving, implementing, and controlling operating procedures as follows:

  - The applicant considers the ISA in identifying needed procedures.

- The procedure specifies operating limits and IROFS.

- Procedures include required actions for off-normal conditions of operation, as well as normal operations.

- If needed, procedures identify safety checkpoints, as appropriate.

- The applicant uses field tests to validate procedures.

- The management personnel who are responsible and accountable for the operation approve the procedures.

- The applicant specifies a mechanism for revising and reissuing procedures in a controlled manner.

- QA elements and CM functions at the facility provide reasonable assurance that current procedures are available and used at all work locations.

- The training program instructs the required personnel in the use of the latest procedures.

- Procedures should incorporate the following elements:

  - title and identifying information, such as number, revision, and date
  - statement of applicability and purpose
  - prerequisites
  - precautions (including warnings, cautions, and notes)
  - important human actions
  - limitations and actions
  - acceptance criteria
  - checkoff lists
  - reference material

- Maintenance procedures involving IROFS commit to the topics listed below for corrective and preventive maintenance and functional testing after maintenance and surveillance activities:

  - Premaintenance activities involve reviews of the work to be performed, including procedure reviews for accuracy and completeness.

  - Steps require notification of all affected parties (operators and supervisors) before performance of work and on completion of maintenance work. The discussion includes potential degradation of IROFS during the planned maintenance.

  - Control of work is ensured by comprehensive procedures to be followed by maintenance technicians. The various safety disciplines, including criticality, fire, radiation, industrial, and chemical process safety, review maintenance procedures. The procedures describe the following:

- ◦ qualifications of personnel authorized to perform the maintenance or surveillance

- ◦ controls on and specification of any replacement components or materials to be used (should be controlled by the CM function to ensure like-kind replacement and adherence to 10 CFR Part 21, "Reporting of Defects and Noncompliance")

- ◦ postmaintenance testing to verify operability of the equipment

- ◦ tracking and records management of maintenance activities

- ◦ safe work practices (e.g., moderation control or exclusion area; radiation or hot work permits; and criticality, fire, chemical, and environmental issues)

- The applicant has formal requirements governing the use of temporary procedures. Temporary procedures may be issued only when permanent procedures do not exist to (1) direct operations during testing, maintenance, and modifications, (2) provide guidance in unusual situations not within the scope of permanent procedures, and (3) provide assurance of orderly and uniform operations for short periods when the facility, system, or component is performing in a manner not covered by permanent procedures. The discussion establishes a timeframe for use of the temporary procedure and sets the same level of review and approval as for permanent procedures.

- The applicant verifies that the procedures are technically accurate and can be performed as written. The applicant periodically reviews the procedures to ensure their continued accuracy and usefulness and establishes the timeframe for reviews of the various types of procedures.

- The applicant describes the use and control of procedures. Provisions allow for operations to stop and place the process in a safe condition if a step of a procedure cannot be performed as written.

- The applicant reviews procedures after unusual incidents, such as an accident, unexpected transient, significant operator error, or equipment malfunction, or after any modification to a system and revises procedures as needed.

- The applicant need not control program and administrative procedures and other nonoperational procedures that do not impact IROFS or other environmental, safety, and health concerns with the stringency applied to operating procedures or management control procedures associated with IROFS specified by the ISA Summary. The applicant should specify the applicability of less stringent procedure control to avoid misunderstandings in implementation.

### 11.4.3.5 Audits and Assessments

The NRC reviewers should find the application acceptable in terms of audits and assessments if it provides reasonable assurance that the following are adequately addressed and satisfied:

- The application describes program directives covering the audit and assessment function (i.e., the activities to be audited, audit frequency, guidance in conducting the audit or assessment, assigned responsibilities for each phase of the work, and procedures for recording the results and recommending actions to be taken).

- The application contains a commitment to conduct internal audits and independent assessments of activities significant to facility safety and environmental protection.

- The application states that audits will be performed to verify that operations are being conducted in accordance with regulatory requirements and license commitments.

- The application states that independent assessments will be conducted by offsite groups or individuals not involved in the licensed activity to verify that the health, safety, and environmental compliance functions are effectively achieving their designed purposes.

- The application states that audits and assessments will be conducted for the areas of radiation safety, nuclear criticality safety, chemical safety, fire safety, environmental protection, emergency management, QA, CM, maintenance, training and qualification, procedures, incident investigation, and records management.

- The application states that qualified personnel without direct responsibility for the function and area being audited or assessed will perform the audits and assessments. The application specifies the staff positions and committees responsible for audits and assessments and describes the levels of management to which results are reported. The systems to provide corrective actions are also described.

### 11.4.3.6 Incident Investigations

The applicant's description of its incident investigations activities and commitments in the application will be acceptable if the reviewer finds reasonable assurance of the following:

- The applicant will establish a formal procedure to investigate abnormal events that may occur during operation of the facility to determine their specific or generic root cause(s), generic implications, and risk significance; to recommend corrective actions; and to report to the NRC as required by 10 CFR 70.50, "Reporting Requirements," and 10 CFR 70.74, "Additional Reporting Requirements." Appendix B to this SRP chapter presents guidance regarding the contents of an incident investigation program or procedure.

- The applicant will monitor and document corrective actions through completion and ensure that corrective actions are taken within a reasonable period to resolve findings from abnormal event investigations.

- The applicant will maintain documentation related to abnormal events for the life of the operation so that "lessons learned" may be applied to future operations of the facility. Details of the event sequence will be compared with accident sequences already considered in the ISA, and the ISA Summary will be modified to include evaluation of the risk associated with accidents of the type actually experienced.

## 11.4.3.7 Records Management

The reviewer will find the applicant's records management system acceptable if the application describes the following criteria:

- The applicant prepares, verifies, characterizes, and maintains records.

- The applicant ensures that records are legible, identifiable, and retrievable for their designated lifetimes.

- The applicant categorizes records by relative safety importance to identify record protection and storage needs and to designate the retention period for individual kinds of records.

- The applicant protects records against tampering, theft, loss, unauthorized access, damage, or deterioration while in storage.

- The applicant establishes and documents procedures specifying the requirements and responsibilities for record selection, verification, protection, transmittal, distribution, retention, maintenance, and disposition.

- The applicant implements procedures that (1) assign responsibilities for records management, (2) specify the authority needed for records retention or disposal, (3) specify which records must have controlled access and provide the controls needed, (4) provide for the protection of records from loss, damage, tampering, and theft or during an emergency, and (5) specify procedures for ensuring that the records management system remains effective.

- The applicant puts procedures in place to promptly detect and correct any deficiencies in the records management system or its implementation.

- The applicant must maintain and update records of IROFS failures in accordance with 10 CFR 70.62(a)(3). The applicant must make record revisions necessitated by postfailure investigation conclusions promptly after completion of the investigation.

- For computer codes and computerized data used for activities relied on for safety, as specified in the ISA Summary, the applicant establishes procedure(s) for maintaining readability and usability of older codes and data as computing technology changes. The procedures could include transfer of the older forms of information and codes for older computing equipment to contemporary computing media and equipment.

Appendix C to this SRP chapter lists the types of records that the system should include.

## 11.4.3.8 Other Quality Assurance Elements

To be acceptable, the applicant's QA elements should be structured to apply appropriate measures to IROFS. The NRC staff expects applicant and licensee QA elements to differ based on the purpose and complexity of the facility and processes.

The ISA Summary should identify the IROFS, the degree of their importance to safety, and related activities that are required for safety. An applicant may choose to apply all QA elements at the highest level to all IROFS or may grade the application in proportion to the item's importance to the achievement of safety.

Other QA may include some or all of the following elements:

- Organization—The applicant may describe the organizational structure, functional responsibilities, lines of authority, and lines of communication for control of activities affecting quality. The organization responsible for ensuring that appropriate QA has been established should have sufficient authority, access to work areas, and organizational independence to perform its responsibilities.

- QA Program—The applicant may describe its application of QA elements in the form of a QA program, in which the applicant commits to meet the relevant requirements of applicable standards. The commitment may describe the applicant's graded approach to QA, in which measures are implemented commensurate with an item's importance to safety, or the commitment may describe a QA program applied to all IROFS. The applicant should fully document, plan, implement, and maintain QA elements to provide reasonable assurance that, together with other management measures, IROFS will be available and reliable when needed.

- Design Control—The applicant should define, control, and verify its design controls. The applicant should specify and correctly translate design inputs to design documents. Controlled measures, commensurate with those applied to the original design, should govern the adequacy of design and design changes. (See Sections 11.3, 11.4.3.1, 11.5.1, and 11.6.1 of this SRP for details on CM.)

- Procurement Document Control—Documents associated with the procurement of items and services include or reference the design-bases information and other documentation necessary to ensure adequate quality. To the extent necessary, suppliers must have a QA program commensurate with the quality level of the item or service to be procured.

- Instructions, Procedures, and Drawing Control—The applicant should ensure that activities affecting the quality of IROFS are prescribed by and performed in accordance with documented instructions, procedures, or drawings appropriate for the circumstances and reference appropriate quantitative or qualitative acceptance criteria. (See Sections 11.3, 11.4.3.4, 11.5.4, and 11.6.4 of this SRP for details on procedures.)

- Document Control—The applicant's document control system describes the preparation, issuance, and modification of documents that specify quality requirements or prescribe activities affecting quality. The document control system is controlled in a manner that ensures that authorized personnel review documents and changes thereto for adequacy and approve them for release. (See Sections 11.3, 11.4.3.1, 11.5.1, and 11.6.1 of this SRP for details on CM and Sections 11.3, 11.4.3.4, 11.5.4, and 11.6.4 for details on procedures.)

- Control of Purchased Items—The applicant may describe controls for the procurement of items and services. Descriptive controls of purchased items and services include, as appropriate, source evaluation and selection, source inspection, audit, the examination of items or services upon delivery or completion, mechanisms for control of changes in items or services, commercial-grade item requirements, and control of supplier nonconformance.

- Identification and Control of Items—The applicant establishes controls to ensure that only the correct items are used or installed. The applicant may describe provisions to identify and maintain traceability of items.

- Control of Processes—The applicant establishes controls of processes affecting the safety of IROFS or related services. Qualified personnel using qualified procedures in accordance with specified requirements perform special processes that control activities, such as welding, heat treating, and nondestructive examination.

- Inspection—When inspections are used to verify conformance of an IROFS item or activity, the applicant should specify the characteristics to be inspected and the inspection methods to be used and plan and execute the inspection. The applicant should then document the inspection results. Qualified personnel other than those who performed or directly supervised the work being inspected should perform the inspections. (See Sections 11.3, 11.4.3.4, 11.5.4, and 11.6.4 of this SRP for details on procedures and Sections 11.3, 11.4.3.3, 11.5.3, and 11.6.3 for details on training and qualification.)

- Test Control—The applicant should conduct tests performed to verify conformance of an IROFS or computer program to specified requirements and demonstrate availability and reliability of performance. The applicant should specify the characteristics to be tested and test methods to be used. Test results should be documented and evaluated against the test requirements and acceptance criteria. (See Sections 11.3.1, 11.4.3.4, 11.5.4, and 11.6.4 of this SRP for details on procedures and Sections 11.3, 11.4.3.3, 11.5.3, and 11.6.3 for details on training and qualification.)

- Control of Measuring and Test Equipment—The applicant should establish controls for tools, gauges, instruments, and other measuring and test equipment used for IROFS and activities affecting IROFS. Controls of measuring and test equipment should consider methods and frequency of calibration, and the applicant should adjust such controls to maintain accuracy within specified limits.

- Handling, Storage, and Shipping—The applicant should consider methods to ensure that handling, storage, cleaning, packaging, shipping, and preservation of IROFS are controlled to prevent damage or loss and to minimize deterioration.

- Inspection, Test, and Operating Status—The applicant should identify the status of inspection and test activities for IROFS, either in the item or in documents traceable to IROFS. The applicant should specify the use of status-indicating devices such as tags, markings, shop travelers, stamps, and inspection records. The applicant should establish provisions to ensure that required inspections and tests are performed and ensure that items that have not passed the required inspections and tests are not inadvertently installed, used, or operated.

- Control of Nonconforming Items—The applicant should describe provisions that specify when IROFS do not conform to specified requirements. The applicant should control items that do not conform to prevent inadvertent installation or use of nonconforming material, parts, equipment, or services. The applicant should specify provisions for identification, documentation, evaluation, segregation, and disposition of nonconforming IROFS and for appropriate notification to affected organizations.

- Corrective Action—The applicant should specify provisions for promptly identifying conditions adverse to quality and correcting them as soon as practicable. (See Sections 11.3, 11.4.3.5, 11.5.5, and 11.6.5 of this SRP for details on audits and assessments and Sections 11.3, 11.4.3.6, 11.5.6, and 11.6.6 for details incident investigations.)

- Quality Assurance Records—QA records and records management systems may be used in lieu of or in conjunction with each other. In either case, the applicant should describe the methods used to document, prepare, maintain, and manage records. The applicant should describe the methods used to protect records against damage, deterioration, or loss. In addition, the applicant should establish and document the requirements and responsibilities for record transmittal, distribution, retention, maintenance, and disposition. (See Sections 11.3, 11.4.3.7, 11.5.7, and 11.6.7 of this SRP for details on records management.)

- Audits—The applicant should plan and schedule audits and assessments to verify compliance with, and to determine the effectiveness of, its QA elements. The applicant should identify responsibilities and procedures for assessing, auditing, documenting, and reviewing results. (See Sections 11.3, 11.4.3.5, 11.5.5, and 11.6.5 of this SRP for details on audits and assessments.)

## 11.5 Review Procedures

For each area of review specified in Section 11.3, the review procedure is identified below. These review procedures are based on the identified SRP acceptance criteria. For deviations from these specific acceptance criteria, the staff should review the applicant's evaluation of how the proposed alternatives to the SRP criteria provide an acceptable method of complying with the relevant NRC requirements identified in Section 11.4.

During the review of the license application and ISA Summary for a planned facility, the reviewer should identify and note any items or issues that should be inspected during an operational readiness review, if such a review will be performed. These items could include confirming that the engineered controls are implemented through procedures and operator training.

If, during the review, the primary reviewer determines a need for additional information, the primary reviewer coordinates a request for additional information with the licensing project manager. The reviewer should ascertain that the approach to management measures is consistent throughout the application.

For an existing facility, the reviewer may consult NRC inspectors to identify and resolve any issues related to the licensing review. For a planned facility, the reviewer may wish to consult with the facility design team to gain a better understanding of the process, its potential hazards, and safety approaches. The reviewer should coordinate these interactions through the licensing project manager.

The primary reviewer will prepare safety evaluation report (SER) input for the licensing project manager in support of the licensing action.

### 11.5.1 Configuration Management

The reviewer should evaluate the six areas of CM described in the next sections.

#### 11.5.1.1 Configuration Management Program

The reviewer should consider whether the CM plan acceptably states management commitments, gives the program directive, and defines key responsibilities, terminology, and equipment scope.

The reviewer should determine whether the applicant's description of overall CM functions covers the following topics: (1) the scope of the IROFS and management measures to be included (coordinating with the reviewer of Chapter 3 of this SRP as necessary), (2) the description and objectives of each CM activity, and (3) the organizational structure and staffing interfaces.

The reviewer should determine that IROFS identified in the ISA Summary are subject to the CM function.

The reviewer should check for appropriate interfaces both within the CM function and with external organizations and functions. In particular, the review should examine functional interfaces with QA, maintenance, and training (including qualification).

The reviewer should look for the applicant's identification of necessary databases and the rules for their maintenance.

#### 11.5.1.2 Design Control Requirements

The reviewer should confirm that the design process leading to drawings and other statements of requirements proceeds logically from the design basis. The design basis is a set of facts about the systems covered by CM which an appropriate authority within the organization has reviewed and approved.

The reviewer should verify that specific personnel are assigned the responsibility for maintaining the design bases and requirements.

The reviewer should verify that the requirements documents clearly define the IROFS to be listed under CM, along with the assignment of any grades or quality levels. The reviewer should coordinate this part of the review with the ISA primary reviewer.

Note that the reviewer, in conjunction with the appropriate technical reviewers, is responsible for determining the adequacy of the reduced levels the applicant would apply to IROFS for accident sequences with lesser consequences.

#### 11.5.1.3 Document Control

The reviewer should evaluate the application to determine whether the CM system captures documents that are relevant and important to safety. These documents should include the design requirements, the ISA, the ISA Summary, as-built drawings, specifications, all operating procedures important to safety, procedures involving training, maintenance, audits and assessments, emergency operating procedures, emergency response plans, system modification documents, assessment reports, and other documents that the applicant deems pertinent to the CM function.

The reviewer should examine information describing a controlled document database used to control documents and track document change status.

The reviewer should confirm that rules of storage for originals or master copies of documents within the scope of the CM function follow the guidance of records management.

*11.5.1.4 Change Control*

The reviewer should verify that the description of change control within the CM function commits to acceptable methods for (1) the identification of changes in configurations that are IROFS, (2) technical and management review of changes, and (3) tracking and implementing changes, including placement of documentation in a document control center and dissemination to affected functions such as training, engineering, operations, maintenance, and other QA elements.

*11.5.1.5 Assessments*

The reviewer should verify that both document assessments and physical assessments (system walk-downs) will be conducted periodically to check the adequacy of the CM functions. The reviewer should also confirm that the applicant will document all assessments and followups.

*11.5.1.6 Design Reconstitution (Existing Facilities Only)*

Design reconstitution may be necessary for existing facilities if current design information is not adequate.

The reviewer should examine the applicant's description of work to establish, organize, and document design requirements and design bases for items for which design information was not available before the application was submitted. This includes the methods used to evaluate, verify, and validate reconstituted design data for IROFS.

The reviewer will seek evidence that the applicant (1) investigated the need for design-bases reconstitution, (2) accomplished reconstitution as necessary, and (3) properly incorporated the new or revised documentation into the CM function.

**11.5.2 Maintenance**

The reviewer will evaluate the applicant's description of how the maintenance function will coordinate with and use the other management measures listed in this chapter. The primary reviewer should consult with supporting reviewers to identify common weaknesses in the applicant's approach and consider these in the review.

### 11.5.3 Training and Qualification

Recognizing that the training objectives and methods and the required personnel qualifications may be graded to correspond to the hazard potential of the facility, the reviewer performs a safety evaluation against the acceptance criteria described in Section 11.4. In particular, the review should accomplish the following:

- The review should evaluate the adequacy of training and qualification on the basis of how well it fulfills the applicant's training objectives, especially when human factors are relied on for safety.

- The review should determine whether the applicant has adequately planned for the training and personnel qualification to be accomplished and whether necessary policies, procedures, and instructions will be in place and appropriate training and qualification will be accomplished before personnel begin activities relied on for safety.

- The reviewer should focus on the training and qualification of personnel who will perform activities relied on for safety.

- The supporting reviewers should become familiar with the applicant's personnel training and qualification commitments and determine whether ongoing activities correspond to them.

- The review should determine whether there is reasonable assurance that the applicant's personnel training and qualification will result in only properly trained and qualified personnel performing activities relied on for safety.

### 11.5.4 Procedures

The reviewer will evaluate whether the applicant has adequately addressed the acceptance criteria listed in Section 11.4. The reviewer will document in an SER that the applicant has committed to the following:

- The applicant includes a statement to follow approved procedures while processing licensed SNM.

- Procedures important to safety are independently verified and validated before use, and this is documented in a program on procedures.

- Procedures exist for the notification of operations personnel before and after maintenance is performed on IROFS, and procedures are in place to control activities.

- An independent, multidisciplinary safety review team reviews and approves changes to operating, management measure, or maintenance procedures controlled by the CM function.

## 11.5.5 Audits and Assessments

The reviewer will determine whether the applicant has adequately planned for audits and assessments to be accomplished and whether necessary programs, personnel, and procedures will be established.

If the applicant refers to other sections of the application when describing its audits and assessments, the reviewer should examine these other sections of the application to determine the applicant's overall commitment to audits and assessments and the proposed method for implementation. The reviewer should confirm that the applicant's audit and assessment commitments are consistent with other sections of the submittal.

## 11.5.6 Incident Investigations

The reviewer will verify that the applicant has described a comprehensive incident investigation function based on the areas of review in Section 11.3 and the acceptance criteria in Section 11.4 of this SRP. For existing facilities, the reviewer should consult with the NRC inspection staff and review any historical information regarding the adequacy of the applicant's incident investigation process.

## 11.5.7 Records Management

The review should determine whether the applicant has adequately implemented a records management system. For fuel cycle facilities that are parts of larger organizations, certain documents may be retained or stored at a site other than the facility site. For example, master drawings for structures might be kept in the engineering department of the headquarters of the parent company. The reviewer may choose to review the physical characteristics of these offsite record storage areas, particularly the storage areas for records related to IROFS for high-consequence accident sequences.

## 11.5.8 Other Quality Assurance Elements

The reviewer should evaluate the applicant's submittal with regard to QA elements against the acceptance criteria in Section 11.4. Supporting reviewers should determine whether IROFS within their areas of review are specified to be within the appropriate QA elements and level. The reviewer should measure the effectiveness of the QA elements design, rather than just verifying the existence of appropriate QA elements.

The reviewer will document in the SER the results of the following:

- The reviewer should determine whether there is reasonable assurance that the applicant's QA elements, maintenance, and CM are coordinated and that the QA elements are an integral part of everyday work activities.

- The reviewer should determine whether there is reasonable assurance that the applicant will be able to monitor the effectiveness of the implementation of QA elements and will make needed adjustments promptly.

- The reviewer should determine that the applicant has specified the QA elements criteria, the basis for choosing the criteria, and the proposed method for implementation.

- If the applicant refers to other sections of the application when describing its QA elements, the reviewer should examine these sections to determine the applicant's commitment to the QA elements and the proposed method for implementation.

## 11.6 Evaluation Findings

The staff's evaluation should verify that the license application provides sufficient information to satisfy the regulatory requirements of Section 11.4.1 of this SRP and that the applicant has appropriately considered the regulatory acceptance criteria in Section 11.4.3 in satisfying the requirements. On the basis of this information, the staff should conclude that this evaluation is complete. The reviewers should write material suitable for inclusion in the SER prepared for the entire application. The SER should include a summary statement of what was evaluated and the basis for the reviewers' conclusions.

In cases in which the SER is drafted in advance of resolving all open issues, the reviewer should document the review as described below and include a list of open issues that require resolution before the staff can reach a conclusion about reasonable assurance of safety. For partial reviews, license revisions, and process changes, the reviewer should use applicable sections of the acceptance criteria and document in the SER only those portions of the submittal that were reviewed.

The following sections present examples of staff documentation for the SER.

### 11.6.1 Configuration Management

The staff has reviewed the CM function for [name of facility] according to Chapter 11 of the SRP (NUREG-1520). [Insert a summary statement of what was evaluated and why the reviewer finds the submittal acceptable.]

The applicant has suitably and acceptably described its commitment to a proposed CM system, including the method for managing changes in procedures, facilities, activities, and equipment for IROFS. Management-level policies and procedures, including an analysis and independent safety review of any proposed activity involving IROFS, are described and will provide reasonable assurance that consistency among design requirements, physical configuration, and facility documentation is maintained as part of a new activity or change in an existing activity involving licensed material. The management measures will include (or do include) the following elements of CM:

1.    Configuration Management

The applicant has put in place or committed to the organizational structure, procedures, and responsibilities necessary to implement CM.

2.    Design Control Requirements

The applicant has documented, and supported by analysis, design requirements and bases. Furthermore, the applicant has ensured that the documentation remains current.

3.   Document Control

The applicant has stored documents, including drawings, in an appropriate and accessible manner. Drawings and related documents captured by the system are those necessary and sufficient to adequately describe IROFS.

4.   Change Control

Responsibilities and procedures adequately describe how the applicant will achieve and maintain strict consistency among the design requirements, the physical configuration, and the facility documentation. The applicant has put in place methods for suitable analysis, review, approval, and implementation of identified changes to IROFS. This includes appropriate CM controls to ensure configuration verification, functional tests, and accurate documentation for equipment or procedures that have been modified.

5.   Assessments

The applicant has committed to an adequate function that includes both initial and periodic assessments, as described in the acceptance criteria in the SRP. The assessments are expected to verify and ensure the adequacy of the CM function.

6.   Design Reconstitution (Existing Facilities Only)

The applicant has adequately described the design reconstitution performed. Current design bases are available and verified for all IROFS, such that the configuration is consistent with the as-built facility documentation.

## 11.6.2 Maintenance

The applicant has committed to maintenance of IROFS. The applicant's maintenance commitments contain the basic elements to maintain availability and reliability: corrective maintenance, PM, functional testing, equipment calibration, and work control for maintenance of IROFS. The applicant's maintenance function is proactive, using maintenance records, PM records, and surveillance tests to analyze equipment performance and to seek the root causes of repetitive failures.

The surveillance and monitoring, PM, and functional testing activities described in the license application provide reasonable assurance that the IROFS identified in the ISA summary will be available and reliable to prevent or mitigate accident consequences.

The maintenance function (1) is based on approved procedures, (2) employs work control methods that properly consider personnel safety, awareness of facility operating groups, QA, and the rules of CM, (3) uses the ISA Summary to identify IROFS that require maintenance and determine the level of maintenance needed, (4) justifies the PM intervals in terms of the equipment reliability goals, (5) provides for training that emphasizes the importance of IROFS identified in the ISA Summary, regulations, codes, and personnel safety, and (6) creates documentation that includes records of all surveillance, inspections, equipment failures, repairs, and replacements of IROFS.

The staff concludes that the applicant's maintenance functions meet the requirements of 10 CFR Part 70 and provide reasonable assurance of public health and safety and the protection of the environment.

### 11.6.3 Training and Qualification

Based on its review of the license application [insert a summary statement of what was evaluated and why the reviewer finds the submittal acceptable], the NRC staff concludes that the applicant has adequately described and assessed its personnel training and qualification in a manner that satisfies the regulatory requirements and is consistent with the guidance in this SRP.

Reasonable assurance exists that implementation of the described training and qualification will result in personnel who are qualified and competent to design, construct, start up, operate, maintain, modify, and decommission the facility safely. The staff concludes that the applicant's plan for personnel training and qualification meets the requirements of 10 CFR Part 70.

### 11.6.4 Procedures

The application describes a suitably detailed process for the development, approval, and implementation of procedures. It has addressed IROFS, as well as items important to the health of facility workers and the public and to the protection of the environment. The staff concludes that the applicant's plan for procedures meets the requirements of 10 CFR Part 70.

### 11.6.5 Audits and Assessments

Based on its review of the license application [insert a summary statement of what was evaluated and why the reviewer finds the submittal acceptable], the NRC staff concludes that the applicant has adequately described its audits and assessments. The staff has reviewed the applicant's plan for audits and assessments and finds it acceptable.

The staff concludes that the applicant's plan for audits and assessments meets the requirements of 10 CFR Part 70 and provides reasonable assurance of protection of the health and safety of the public, workers, and the environment.

### 11.6.6 Incident Investigations

The applicant has committed to and established an organization responsible for (1) performing incident investigations of abnormal events that may occur during operation of the facility, (2) determining the root cause(s) and generic implications of the event, and (3) recommending corrective actions for ensuring a safe facility and safe facility operations, in accordance with the acceptance criteria of Section 11.4 of the SRP.

The applicant has committed to monitoring and documenting corrective actions through to completion.

The applicant has committed to the maintenance of documentation so that "lessons learned" may be applied to future operations of the facility.

Accordingly, the staff concludes that the applicant's description of the incident investigation process complies with applicable NRC regulations and is adequate.

### 11.6.7 Records Management

The staff has reviewed the applicant's records management system against the acceptance criteria and concludes that the system (1) will be effective in collecting, verifying, protecting, and storing information about the facility and its design, operations, and maintenance and will be able to retrieve the information in readable form for the designated lifetimes of the records, (2) will provide a records storage area(s) with the capability to protect and preserve health and safety records that are stored there during the mandated periods, including protection of the stored records against loss, theft, tampering, or damage during and after emergencies, and (3) will provide reasonable assurance that any deficiencies in the records management system or its implementation will be detected and corrected promptly.

### 11.6.8 Other Quality Assurance Elements

The SER should include a summary statement of what the NRC evaluated and the basis for the reviewer's conclusions. The review should demonstrate the adequacy of the applicant's use of other QA elements, as applied to IROFS, for design, construction, and operations. The SER should include statements like the following:

The NRC staff concludes that the applicant has adequately described the application of other QA elements (and the applicable QA elements of its principal contractors). The staff also concludes the following:

- The applicant has established and documented a commitment to an organization responsible for developing, implementing, and assessing the management measures for providing reasonable assurance of safe facility operations, in accordance with the criteria in Section 11.4 of the SRP.

- The applicant has established and documented a commitment to QA elements, and the administrative measures for staffing, evaluating performance, assessing findings, and implementing corrective action are in place.

- The applicant has developed a process for preparation and control of written administrative plant procedures, including procedures for evaluating changes to procedures, IROFS, and tests. The applicant has committed to implement and maintain a process for review, approval, and documentation of procedures.

- The applicant has established and documented surveillances, tests, and inspections to provide reasonable assurance of satisfactory in-service performance of IROFS.

- The applicant will ensure that periodic independent audits are conducted to determine the effectiveness of the management measures. Management measures will provide for documentation of audit findings and implementation of corrective actions.

- The applicant has established and documented training requirements to provide employees with the skills to perform their jobs safely. The applicant has also provided management measures for the evaluation of the effectiveness of training against predetermined objectives and criteria.

- The organizations and persons performing QA element functions have the required independence and authority to effectively carry out their QA element functions without undue influence from those directly responsible for process operations.

- QA elements cover the IROFS, as identified in the ISA Summary, and the applicant has established measures to prevent hazards from becoming pathways to higher risks and accidents.

Accordingly, the staff concludes that the applicant's use of other QA elements meets the requirements of 10 CFR Part 70 and provides reasonable assurance that public health and safety and the environment are protected.

## 11.7 References

American National Standards Institute/American Society of Mechanical Engineers Standard, "Quality Assurance Requirements for Nuclear Facility Applications," ANSI/ASME NQA-1-1994.

*U.S. Code of Federal Regulations*, Chapter I, Title 10, "Energy," Part 50, "Domestic Licensing of Production and Utilization Facilities."

*U.S. Code of Federal Regulations*, Chapter I, Title 10, "Energy," Part 70, "Domestic Licensing of Special Nuclear Material."

*U.S. Code of Federal Regulations*, Chapter I, Title 10, "Energy," Part 21, "Reporting of Defects and Noncompliance."

U.S. Nuclear Regulatory Commission, "Guidance on Management Controls/Quality Assurance, Requirements for Operation, Chemical Safety, and Fire Protection for Fuel Cycle Facilities," *Federal Register,* Vol. 54, No. 53, pp. 11590–11598, March 21, 1989.

U.S. Nuclear Regulatory Commission, "Suggested Guidance Relating to Development and Implementation of Corrective Action," Information Notice 96-28, May 1966.

U.S. Nuclear Regulatory Commission, "Human Factors Engineering Program Review Model," NUREG-0711, Revision 2, February 2004.

# APPENDIX A

# CHECKLIST FOR PROCEDURES

Written procedures should cover all activities listed below. This list is not intended to be all-inclusive or to imply that procedures must be developed with the same titles as those on the list.

1.  Management Control Procedures

    - training
    - audits and assessments
    - incident investigation
    - records management
    - configuration management
    - quality assurance
    - equipment control (lockout/tagout)
    - shift turnover
    - work control
    - procedure management
    - nuclear criticality safety
    - fire protection
    - radiation protection
    - radioactive waste management
    - maintenance
    - environmental protection
    - chemical process safety
    - operations
    - calibration control
    - preventive maintenance

2.  Operating Procedures

    - system procedures that address startup, operation, shutdown, control of process operations, and recovery after a process upset

        – ventilation
        – criticality alarms
        – shift routines, shift turnover, and operating practices
        – decontamination operations
        – uranium recovery
        – facility utilities (air, other gases, cooling water, fire water, steam)
        – temporary changes in operating procedures

    - abnormal operation/alarm response

        – loss of cooling water
        – loss of instrument air
        – loss of electrical power
        – loss of criticality alarm system

         –      fires
         –      chemical process releases

3.     Maintenance Activities That Address System Repair, Calibration, Surveillance, and Functional Testing

- repairs and preventive repairs of items relied on for safety (IROFS)
- testing of criticality alarm units
- calibration of IROFS
- high-efficiency particulate air filter maintenance
- functional testing of IROFS
- relief valve replacement/testing
- surveillance/monitoring
- pressure vessel testing
- nonfired pressure vessel testing
- piping integrity testing
- containment device testing

4.     Emergency Procedures

- response to a criticality
- hazardous process chemical releases (including uranium hexafluoride)

# APPENDIX B

## INCIDENT INVESTIGATION PROGRAMS AND PROCEDURES

The following eight items are good practices to incorporate in incident investigation programs or procedures or both:

(1) The investigation of an abnormal event should begin as soon as possible after the event has been brought under control.

(2) The incident investigation program contains a documented procedure for investigating an abnormal event. This procedure is separate from any required emergency plan.

(3) The program includes a description of the functions, qualifications, and responsibilities of the manager who would lead the investigative team and those of the other team members; the scope of the team's authority and responsibilities; and an assurance of management cooperation.

(4) Qualified internal or external investigators are appointed to serve on investigating teams when required. The teams should include at least one process expert and at least one team member trained in root cause analysis.

(5) The program contains guidance for personnel conducting the investigation on how to apply a reasonable, systematic, structured approach to determine the specific or generic root cause(s) and generic implications of the problem. The level of investigation should be based on a graded approach relative to the severity of the incident.

(6) The incident investigation team has assurance of the team's authority to obtain all information considered necessary and is independent from the functional area involved in the incident under investigation.

(7) The investigation process and investigating teams are independent of the line management.

(8) Auditable records and documentation related to abnormal events, investigations, and root cause analysis are maintained. For each abnormal event, the incident report should include a description, contributing factors, a root cause analysis, and findings and recommendations. Relevant findings are reviewed with all affected personnel.

# APPENDIX C

# RECORDS

The requirements for records management vary according to the nature of the facility and the hazards and risks it poses. Examples of the records required by Title 10 of the *Code of Federal Regulations* (10 CFR) Part 19, "Notices, Instructions and Reports to Workers: Inspection and Investigations"; 10 CFR Part 20, "Standards for Protection against Radiation"; 10 CFR Part 21, "Reporting of Defects and Noncompliance"; 10 CFR Part 25, "Access Authorization"; and 10 CFR Part 70, "Domestic Licensing of Special Nuclear Material," are listed below. The records are listed under the chapter headings of the Standard Review Plan (SRP). The list is not intended to be exhaustive or prescriptive. Different or additional records may be required in certain circumstances. The applicant may also choose to organize the records in other ways.

## Examples of Records

**SRP Chapter**

1.  General Information

    -   construction records

    -   facility and equipment descriptions and drawings

    -   design criteria, requirements, and bases for items relied on for safety (IROFS), as specified by the facility configuration management (CM) function

    -   records of facility changes and associated integrated safety analyses, as specified by the facility CM function

    -   safety analyses, reports, and assessments

    -   records of site characterization measurements and data

    -   records pertaining to onsite disposal of radioactive or mixed wastes in surface landfills

    -   procurement records, including specifications for IROFS

2.  Organization and Administration

    -   administrative procedures with safety implications

    -   change control records for material control and accounting program

    -   organization charts, position descriptions, and qualification records

    -   safety and health compliance records, medical records, personnel exposure records

- quality assurance records

- safety inspections, audits, assessments, and investigations

- safety statistics and trends

3.  Integrated Safety Analysis

4.  Radiation Safety

- bioassay data
- exposure records
- radiation protection (and contamination control) records
- radiation training records
- radiation work permits

5.  Nuclear Criticality Safety

- nuclear criticality control written procedures and statistics

- nuclear criticality safety analyses

- records pertaining to nuclear criticality inspections, audits, investigations, and assessments

- records pertaining to nuclear criticality incidents, unusual occurrences, or accidents

- records pertaining to nuclear criticality safety analyses

6.  Chemical Safety

- chemical process safety procedures and plans

- records pertaining to chemical process inspections, audits, investigations, and assessments

- diagrams, charts, and drawings

- records pertaining to chemical process incidents, unusual occurrences, or accidents

- chemical process safety reports and analyses

- chemical process safety training

7.  Fire Safety

- fire hazard analysis

- fire prevention measures, including hot-work permits and fire watch records

- records pertaining to inspection, maintenance, and testing of fire protection equipment

- records pertaining to fire protection training and retraining of response teams

- prefire emergency plans

8.    Emergency Management

- emergency plan(s) and procedures

- comments on emergency plan from outside emergency response organizations

- emergency drill records

- memoranda of understanding with outside emergency response organizations

- records of actual events

- records pertaining to the training and retraining of personnel involved in emergency preparedness functions

- records pertaining to the inspection and maintenance of emergency response equipment and supplies

9.    Environmental Protection

- environmental release and monitoring records
- environmental report and supplements to the environmental report, as applicable

10.    Decommissioning

- decommissioning records
- financial assurance documents
- decommissioning cost estimates
- site characterization data
- final survey data
- decommissioning procedures

11.    Management Measures

11.1    Configuration Management

- safety analyses, reports, and assessments that support the physical configuration of process designs and changes to those designs

- validation records for computer software used for safety analysis or material control and accounting

- integrated safety analysis documents, including process descriptions, plant drawings and specifications, and purchase specifications for IROFS

- approved current operating procedures and emergency operating procedures

## 11.2 Maintenance

- record of IROFS failures (required by 10 CFR 70.62, "Safety Program and Integrated Safety Analysis")

- preventive maintenance records, including trending and root cause analysis

- calibration and testing data for IROFS

- corrective maintenance records

## 11.3 Training and Qualification

- personnel training and qualification records
- procedures

## 11.4 Procedures

- standard operating procedures
- functional test procedures

## 11.5 Audits and Assessments

- audits and assessments of safety and environmental activities

## 11.6 Incident Investigations

- investigation reports
- changes recommended by investigation reports and how and when implemented
- summary of reportable events for the term of the license
- incident investigation policy

## 11.7 Records Management

- policy
- material storage records
- records of receipt, transfer, and disposal of radioactive material

| NRC FORM 335<br>(9-2004)<br>NRCMD 3.7 | U.S. NUCLEAR REGULATORY COMMISSION | 1. REPORT NUMBER<br>(Assigned by NRC, Add Vol., Supp., Rev., and Addendum Numbers, if any.) |
|---|---|---|
| **BIBLIOGRAPHIC DATA SHEET**<br>*(See instructions on the reverse)* | | NUREG-1520, Rev. 01 |

**2. TITLE AND SUBTITLE**

Standard Review Plan for the Review of a License Application for a Fuel Cycle Facility

**3. DATE REPORT PUBLISHED.**

| MONTH | YEAR |
|---|---|
| May | 2010 |

**4. FIN OR GRANT NUMBER**

**5. AUTHOR(S)**

Division of Fuel Cycle Safety and Safeguards

**6. TYPE OF REPORT**

**7. PERIOD COVERED** *(Inclusive Dates)*

**8. PERFORMING ORGANIZATION - NAME AND ADDRESS** *(If NRC, provide Division, Office or Region, U.S. Nuclear Regulatory Commission, and mailing address; if contractor, provide name and mailing address.)*

Division of Fuel Cycle Safety and Safeguards

Office of Nuclear Material Safety and Safeguards

U.S. Nuclear Regulatory Commission

Washington DC 20555-0001

**9. SPONSORING ORGANIZATION - NAME AND ADDRESS** *(If NRC, type "Same as above"; if contractor, provide NRC Division, Office or Region, U.S. Nuclear Regulatory Commission, and mailing address.)*

Same as above

**10. SUPPLEMENTARY NOTES**

**11. ABSTRACT** *(200 words or less)*

This "Standard Review Plan (SRP) for the Review of a License Application for a Fuel Cycle Facility" (NUREG-1520) provides guidance to the staff reviewers in the U.S. Nuclear Regulatory Commission (NRC), Office of Nuclear Material Safety and Safeguards who perform safety and environmental impact reviews of applications to construct or modify and operate nuclear fuel cycle facilities. The SRP is intended to be a comprehensive and integrated document that provides the reviewer with guidance that describes methods or approaches that the staff has found acceptable for meeting NRC requirements. As such, this SRP ensures the quality, uniformity, and predictability of the staff reviews. This SRP also makes information about licensing acceptance criteria widely available to interested members of the public and the regulated industry and is intended to improve industry and public stakeholder understanding of the staff review process. Each SRP section addresses the responsibilities of the staff reviewers, the matters that they review, the Commission's regulations pertinent to specific technical matters, the acceptance criteria used by the staff, the process and procedures used to accomplish the review, and the conclusions that are appropriate to summarize the review.

This SRP is not a substitute for NRC regulations and compliance is not required. The approaches and methods in this report are provided for information only. Methods and solutions different from those described in this report will be acceptable if they provide a basis for the staff to make the determination needed to issue or continue a license.

**12. KEY WORDS/DESCRIPTORS** *(List words or phrases that will assist researchers in locating the report.)*

Standard Review Plan, SRP, Fuel Cycle, fuel fabrication, safety review, environmental review, NMSS, technical review, FCSS, ISA, ISA Summary, acceptance criteria, uranium, enrichment.

**13. AVAILABILITY STATEMENT**

unlimited

**14. SECURITY CLASSIFICATION**

*(This Page)*

unclassified

*(This Report)*

unclassified

**15. NUMBER OF PAGES**

**16. PRICE**

Printed
on recycled
paper

Federal Recycling Program